HAZARDOUS MATERIAL AND HAZARDOUS WASTE

A Construction Reference Manual

Francis J. Hopcroft, P.E.
David L. Vitale, M.Ed.
Donald L. Anglehart, Esq.

Illustrated by Carl W. Linde

HAZARDOUS MATERIAL AND HAZARDOUS WASTE

A Construction Reference Manual

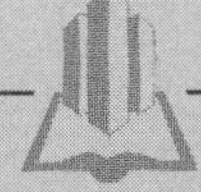

Francis J. Hopcroft, P.E.
David L. Vitale, M.Ed.
Donald L. Anglehart, Esq.

R.S. MEANS COMPANY, INC.
CONSTRUCTION CONSULTANTS & PUBLISHERS
100 Construction Plaza
P.O. Box 800
Kingston, Massachusetts 02364-0800
(617) 585-7880

This book was edited by Mary Greene and Ernest Williams. Typesetting was supervised by Joan Marshman. The book and jacket were designed by Norman Forgit. Illustrations by Carl Linde.

Printed in the United States of America

10 9 8 7 6 5

Library of Congress Catalog Card Number 89-189295

ISBN 0-87629-136-1

Foreword

Proper handling of hazardous materials and hazardous wastes is an issue that contractors can no longer afford to ignore. Gone are the days when all debris from a construction project could be tossed into a dumpster and hauled to the nearest landfill, with no thought given to health or legal repercussions.

Today's federal and state laws specifically define the substances that threaten human health and/or the environment. Contractors who fail to comply with the legal standards for handling these substances expose themselves to tremendous liability in the form of substantial government fines and suspension of work, as well as civil suits from injured workers and property owners whose land has been contaminated. Beyond the financial implications are the moral issues of protecting workers from serious injury and preventing long-term damage to the environment.

The purpose of this book is to provide contractors with a source for specific, practical, step-by-step actions they can take to ensure project safety and compliance with regulations.

The treatment of hazardous materials is the subject of Part I. The first two chapters introduce the basic requirements of the Hazard Communication Standard issued by OSHA (Occupational Safety and Health Administration). Chapter 3 goes on to outline the Hazard Communication Program, an information and training program which the contractor is legally obligated to carry out.

The next four chapters address the issues of handling hazardous materials—from storage and inventory requirements to spill prevention and cleanup. An entire chapter is devoted to the special considerations of asbestos materials.

Hazardous waste—those substances that are left over from the hazardous materials used in the construction of a project—are the focus of Part II. Chapter 8 is an overview of the Resource Conservation and Recovery Act (RCRA). The remainder of Part II, Chapters 9-15, cover storage, handling, and measures to reduce the quantity of hazardous waste. Separate chapters are devoted to asbestos and PCB wastes, unexpected subsurface contamination, and leaking underground storage tanks.

The appendices follow, with lists of EPA (Environmental Protection Agency), DOT (Department of Transportation), and industry codes, and the addresses of state hazardous waste management agencies. Finally, a Glossary defines terms useful to anyone implementing or interpreting a hazardous materials and hazardous wastes management plan.

To ensure project safety and avoid liability, contractors must educate themselves and their workers on the subject of the hazardous substances they encounter. By following the guidelines presented in this book, contractors should be able to draw up and carry out an effective program that will not only protect workers and the environment, but will also guard their own businesses against liability and economic disaster.

Editor's Note: *The information in this book has been compiled using sources believed to be reliable and representative of expert opinion on the subjects as of late 1988. The written materials are not intended, however, to represent governmental standards completely, in every detail. Moreoever, the plans and sketches should be modified based on individual site needs, and updated as new information becomes available and new government regulations take effect. While every reasonable effort has been made to provide dependable information, the authors cannot assume responsibility for the consequence of its use.*

Table of Contents

HAZARDOUS MATERIAL AND HAZARDOUS WASTE

Part I

Chapter 1

Standards and Requirements

The effective management of hazardous wastes begins with the management of hazardous materials. All wastes generated on construction sites originate with the materials used; the nature of those materials determines how hazardous the wastes will be. The way the materials are managed determines the way in which the wastes must be managed.

OSHA *Hazard Communication Standard*

Employer Responsibility

The management of hazardous materials in any workplace, including a construction site, is regulated by the Occupational Safety and Health Administration's (OSHA) Hazard Communication Standard, also known as *the Standard*. The Standard requires all employers to:

- Inventory the hazardous materials on the job site
- Obtain a Material Safety Data Sheet (MSDS) for each material
- Post both the inventory list and the MSDS forms in a location accessible to all employees
- Create a Written Hazard Communication Program
- Establish a training program

The MSDS Form is prepared by the manufacturer of the material. It informs the reader of the physical and chemical properties of the material, and any health hazards associated with its use. The MSDS also provides guidance on the personal protective equipment to be used when handling the material. Other information includes a list of symptoms resulting from exposure, and what to do in the event of exposure.

After the inventory and MSDS forms are collected, the employer is required to develop a written **Hazard Communication Program**. The purpose of this program is to inform each employee of the health and safety risks associated with the material(s) that the employee is required to use.

Finally, the employer must provide a **training program** to teach each employee how to read an MSDS, what to do in an emergency, and how to protect him or herself in the workplace. The employer must ensure that each employee is provided with the proper personal protective equipment for the job being done, and that he or she knows how and when to use the equipment.

Employee Responsibility

The Standard also imposes the following responsibilities on the employee.

- To pay attention during the training sessions
- To use the personal protective equipment provided
- To follow appropriate safety guidelines

The employer and employees must work together to ensure safety in the workplace.

Hazardous Materials

Storage

The benefits of even the best hazard communication program can be undermined by the haphazard storage of materials on a project site. Proper storage is a crucial ingredient in project safety. Safe storage requires careful consideration of the following factors:

- The compatibility or incompatibility of the various components
- The temperature at which components or materials become unstable
- The consequences of inadvertent damage to containers or spills of materials

Storage generally must provide segregation, temperature control, protection from moisture and weather, and protection from container damage.

Common sense rules can be used in almost all cases. The common sense, however, comes from a review of the inventory and MSDS forms to identify proper storage conditions for each material.

Corrosive or Flammable Materials

Highly corrosive materials, like acids, should be stored in fire-resistant metal cabinets that have protective coatings and spill containment wells. Highly flammable materials should be stored in similar cabinets. However, *strong acids (oxidizers) should* **never** *be stored in the same cabinet with flammables.*

Materials that have a chemical reaction when joined together must be separated carefully in storage. For example, a liquid acid dripping onto a bag of powdered alkali can cause a rapid generation of heat. A resulting fire could rupture more acid containers and thereby spill more acid on more alkali at an increasing rate. An inferno can rapidly develop.

Similarly, dry alkalis can become very corrosive when wet. Bags piled on the floor, when subjected to moisture seeping in or rainwater dripping through an opening, can cause significant damage before the problem is discovered. Dry corrosive materials should be stored on wooden pallets and covered with plastic.

Handling

Eventually the materials are moved out of storage and incorporated into the work. This is where the quality of the Hazard Communication Program training becomes evident. If the training was conducted properly, and if the employees were paying attention, then the employees will be wearing the appropriate personal protective equipment when the material is moved out of storage, the material will be used in accordance with the MSDS guidelines and the job will be done safely.

Of course, even the most conscientious worker occasionally has an accident. For example, liquids spill while being poured, containers are dropped and broken, or bags may be ripped open at inopportune moments. It is important to anticipate as many of these accidents as possible and to pre-plan where they may occur so as to minimize the associated damage and health risks.

For example, when mixing an acid into water to create a more dilute solution, the mixing container should be set inside a mixing box lined with an acid neutralizer. If a drip or spill occurs, it is trapped in the box, where it is neutralized. The health risks are thereby minimized.

Similarly, when a solvent is stored in a drum on a rack, a drip pan should be placed below the drum to catch drips from the spigot. Since spigots are notorious for leaking, it must be assumed that every spigot will leak constantly. Consequently, the size of the drip pans should be adequate to hold the entire contents of the drum in the event the spigot mechancism should fail during the night. Again, these measures limit the problem and minimize the health risks.

Each material has its own associated hazards and must be independently evaluated to ensure proper and safe material handling. A proper material management program requires that the following items be kept available on the job site:

- Proper neutralizing materials for acids
- Absorbents for other liquids
- Containers for spilled solids

The Requirements of RCRA

Whether from having been used properly and having served its purpose, or from cleanup of a spill or accident, eventually all materials not totally consumed in construction will become a waste. Some are harmless or innocuous, such as lumber scraps and cardboard boxes, and are easily disposed to any solid waste disposal facility. Others, however, are as dangerous to humans as the materials from which they came. The disposal of those hazardous wastes is controlled by the **Resource Conservation and Recovery Act (RCRA)**.

RCRA defines anyone who creates a hazardous waste as a *generator*. Three categories of generator are recognized: *Large Quantity Generators* (LQG), *Small Quantity Generators* (SQG), and *Very Small Quantity Generators* (VSQG).

In general, anyone who generates more than 2,200 lbs. of hazardous waste in one month is a Large Quantity Generator. Those who generate more than 220 lbs. per month, but fewer than 2,200 lbs., are Small Quantity Generators. Anyone who generates fewer than 220 lbs. of hazardous waste per month is a Very Small Quantity Generator. Figure 1.1 summarizes the various criteria defining hazardous waste generators.

All generators must register with the Environmental Protective Agency (EPA), as well as with their state environmental protection agency. They must also maintain records documenting the status of wastes they generate, and, except under certain conditions, manifest when the wastes are disposed. The requirements differ only in the amount of paperwork that must be kept to document the quantities of wastes created.

RCRA specifies how drums or containers of waste must be marked, how long containers can be stored on site, and how storage areas should be constructed and arranged. RCRA is a comprehensive and complex law, as are the regulations that implement it. Nevertheless, it is essential that site managers understand this law and its regulations; failure to comply can lead to serious health risks and heavy fines.

Hazardous Waste: A Management Plan

The best way to comply with RCRA throughout the job is to carefully pre-plan the management of hazardous wastes. The completeness of the plan will determine its effectiveness. A complete plan includes the following actions.

Notification

The first step is to notify EPA and the appropriate state agency that hazardous wastes will be generated. The notification form asks for information regarding the nature of the wastes to be generated and the expected quantities. A filer should be careful in estimating these values. A low estimate could cause a generator to be improperly classified as too small, leading to regulatory problems later. On the other hand, if an estimate is improperly classified as too large, the result is unnecessary paperwork and expense. It is generally better, however, to be classified larger than necessary rather than smaller.

Guide to Determining Status and Regulatory Requirements for Hazardous Waste Generators

EPA Status	Hazardous Waste Management Accumulation Limits			Management Requirements		
	Time (Days)	Volume in Tanks (KG)	Volume in Containers (KG)	Accumulation Area Standards	Emergency Preparation	Personnel Training Contingency Plan & Annual Report
LQG	90	No limit	No limit	Yes		Yes
SQG	180	6,000	2,000	Yes	Yes	
VSQG	180	600	600	Yes	Yes	

Definitions:

EPA Status	Kilograms/Month
VSQG	less than 100
SQG	100-999
LQG	1,000 or more

Conversions:

Kilograms (metric)	Gallons (varies by substance)
100	25-27
600	150-165
1,000	250-270
2,000	500-550
6,000	1,500-1,650

NOTES:
1. *This table does not apply to generators of acutely hazardous waste.*
2. *Various state regulations may be more stringent regarding storage times and accumulation limits.*
3. *A VSQG who must transport wastes more than 200 miles may apply for a permit to store wastes for up to 270 days.*

Figure 1.1

Site Audit

The second step is to conduct an audit of the job site or to develop a probable list of hazardous wastes that may be expected on a proposed site. With the audit in hand, an evaluation can be made of the compatibility or incompatibility of various wastes.

Obtaining MSDS Forms

The third step is to gather the Material Safety Data Sheet (MSDS) forms for each known or expected material from which the wastes are generated. This information is used to more carefully evaluate the storage requirements, the personal protective equipment needs, and the waste documentation data to be accumulated.

Ensuring Proper Procedures

Next, the handling, storage, and disposal practices currently used must be evaluated to determine whether they are proper and appropriate. If current procedures are not adequate, standard operating procedures must be developed to address those activities.

The Training Program

A training program for all personnel is conducted next. The training program for hazardous wastes is similar to that provided for the hazardous materials on the site. That means that employees are given information on the hazards posed by the wastes, how to protect themselves from exposure, what to do if exposure occurs, and what to do in the event of a spill or accident.

Finally, with all the other elements in place, the management plan can be implemented.

Handling Hazardous Wastes

Handling hazardous wastes is a serious business that requires the same care and use of personal protective equipment as the handling of materials from which the wastes came. The MSDS form is the best source of guidance on the proper handling of hazardous wastes.

The same procedures that are used to pour, transfer, or handle the raw material also apply to hazardous wastes. An acid still behaves like an acid, even when it is contaminated. Spills and drips from pouring or transfer operations must be quickly contained and cleaned up. The area ventilation and fire protection practices applied to hazardous material transfer and storage should also be applied to hazardous waste.

Storage of Hazardous Wastes

There are safe ways to store hazardous wastes and there are legal limits on the length of time wastes may be stored. Safe storage information can be found on the appropriate MSDS sheets for a particular substance or material. Legal storage times are shown on Figure 1.1. All containers and storage sheds used for hazardous wastes should be fully labeled, in accordance with RCRA regulations.

Separating Incompatibles

As with hazardous materials, hazardous waste storage plans must consider separation and segregation of incompatibles. This is done for two reasons. First, the same kind of chemical reaction can occur with wastes as can occur with raw materials. Those reactions can lead to fires, toxic fumes, or other serious hazards. Second, mixing wastes, even those that may be chemically stable, can lead to significant disposal costs.

Certain wastes may be mixed, such as used motor oils and lubricants, which can be jointly disposed as waste oil. However, gasoline, benzene, toluene, kerosene, and other solvents should not be mixed with the oils because these compounds, even in small quantities, interfere with the safe reclamation of the oils. Consequently, disposing of the mixture becomes more difficult and more expensive.

Metal Drums

Most liquid wastes are stored in 35-gallon or 55-gallon metal drums. These drums are readily available, easily moved when full, and provide spill-proof seals when properly fitted with screw-in bungs. Filling is accomplished either by direct pumping or by pouring liquids from transfer containers through a funnel.

Metal drums are recommended for liquid waste storage, except in the case of strongly acidic wastes, such as used battery acids. Battery acids should be stored in compatible rigid plastic containers such as low density polyethylene, high density polyethylene, polypropylene, or polymethylpentene (TPX).

Storage Area Requirements

Storage areas should be designed in such a way that a catastrophic failure of all containers in the area can be fully contained. This kind of protection normally requires a diked berm built around the storage site. Moreover, the area must be maintained at a temperature above 35°F to prevent freezing liquids from rupturing storage drums, and below 120°F to prevent the build-up of excess pressure inside the drums, which could also lead to a rupture. Note that federal regulations require only that 10% of the total contained volume (or 110% of the volume of the largest container in the storage area) must be containable within the diked area. The more stringent 100% total volume containment is recommended, however, wherever possible.

Solid Waste Storage

Hazardous solid wastes are usually composed of damp absorbents which were used to absorb spills of hazardous liquids, the rags used to apply or clean up solvents, dried paints, and containers that once held hazardous materials. These items are best stored in metal or plastic drums with covers that open completely in order to facilitate filling and dumping. Dry materials, such as plastic sheeting used to contain contaminated soils, and contaminated personal protective equipment, can be stored in paper drums with tightly fitted lids. Materials containing no free liquids do not require a bermed storage area, if the following provisions are met:

1. The actual area must be sloped to prevent water accumulation, and
2. The dry materials must be raised above the floor slightly, as on pallets.

Problems with Dumpster Storage

The use of conventional dumpsters for hazardous waste storage is discouraged due to the high probability that such material may be inadvertently disposed as a non-hazardous waste. In addition, cleaning a dumpster that has been used for hazardous waste storage can be a risky undertaking.

Temperature Requirements

Solid wastes are usually not harmed by exposure to freezing. However, high temperatures (greater than 120°F) can cause spontaneous combustion. Exposure to the elements, particularly in the case of paper drums, can cause rapid container deterioration and leakage.

Unexpected Subsurface Contaminations

One of the biggest and most costly surprises on a construction site is the discovery of hidden, subsurface contamination. After months of careful bidding, planning, and organizing, the contractor mobilizes on the site, all ready to go to work. The first thing his excavator finds is a funny colored soil under the overlying gravel, a pungent odor from the hole, or a pile of leaking drums buried surreptitiously under the surface.

There is no option in that situation except to stop the work until the nature and extent of the problem can be ascertained and a cleanup procedure developed. The choice of cleanup procedure will depend, of course, on the nature of the contaminant found, the concentration in the soil and water, and the extent of the problem.

It must be recognized up-front that every gram of contaminated soil or water is a hazardous waste. Once it is picked up, it cannot be put down again, except in strict accordance to RCRA guidelines. Ignoring the problem will not make it go away, and an improper approach to cleanup can lead to significant future liability.

The topics introduced in this chapter are more fully developed in the following chapters. The reader is encouraged to scan the Table of Contents for the particular topics, as each chapter stands alone as a reference to a special issue.

The management of hazardous wastes on construction sites is an undertaking of monumental importance. Approaching it in a step-by-step manner will make the task easier and the outcome significantly better.

Chapter 2

The OSHA Hazard Communication Standard

In 1970, Congress passed the Occupational Safety and Health Act. That act includes a requirement that any occupational safety or health standard promulgated by the Secretary of Labor must detail the use of labels or other appropriate forms of warning. Such warnings must ensure that employees are apprised of all hazards to which they are exposed in the workplace, the relevant symptoms of exposure, and appropriate emergency treatment in case of exposure, and the proper conditions and precautions for safe use.

The Occupational Safety and Health Administration has estimated that American workers are exposed to as many as 575,000 hazardous chemical products. Because of this enormous number, OSHA has concluded that a substance-by-substance rule-making approach to the transmittal of hazard information is impracticable. Therefore, OSHA has decided to specify regulatory requirements for hazard information transmittal on a generic basis, through its comprehensive *Hazard Communication Standard*. The purpose of the Hazard Communication Standard (known as *the Standard*) is to ensure that the hazards of all chemicals produced in, or imported into, the United Stated are fully evaluated, and that information concerning their hazards is transmitted to employers and employees.

For a number of years, OSHA had included in its regulations for the construction industry certain requirements for training and informing workers of chemical hazards. However, those requirements proved largely ineffective due to their lack of specificity. For example, OSHA's Safety and Health Regulations for Construction, for several years, stated the following general requirement for construction industry employers:

> *Employees required to handle or use poisons, caustics, and other harmful substances shall be instructed regarding safe handling and use, and be made aware of the potential hazards, personal hygiene, and personal protective measures required.*

Unfortunately, until the Standard became applicable to the construction industry on September 23, 1987, the OSHA Safety and Health Regulations for Construction did not specify the mechanisms by which employers were to obtain from suppliers and manufacturers the basic chemical hazard information concerning products used in the industry, nor the means by which that information should be imparted to employees.

Participants in the construction industry were instrumental in developing the comprehensive OSHA Hazard Communication Standard in effect today. The Construction Advisory Committee submitted to the Assistant Secretary of

Labor in 1980 its study entitled "Report on Occupational Health Standards for the Construction Industry." That report included recommendations for labels, material safety data sheets, and training–all the major components of the present Standard. The report noted that construction employers often did not have access to the information necessary to develop appropriate signs, labels, or material safety data sheets, nor were they always aware of the hazards associated with a particular product or device. Many items were not accompanied, upon purchase, by appropriate labels and data sheets.

Although OSHA apparently agreed with many of the observations made by the Construction Advisory Committee in the 1980 report, it did not expand its Hazard Communication Standard to cover construction employers (and other non-manufacturing employers) until forced to do so by a federal appeals court in 1985. At that time, the court ruled that OSHA would be required to apply the Standard – not just to the manufacturing sector to which it initially applied, but to *all* sectors, unless the agency could state persuasive reasons why such application would not be feasible.

As a direct result of the 1985 case, OSHA expanded the scope of the Standard to include the construction industry and other industrial sectors. Figure 2.1 shows estimates of construction worker exposure to hazardous chemicals, as published by OSHA.

Scope and Application

OSHA implements the Standard through a comprehensive program that emphasizes labeling and the dissemination of information. The requirements set forth in the Standard apply to *chemical manufacturers, importers, distributors,* and *employers*. Each of these four entities is particularly identified within the Standard. Chemical manufacturers are defined by the Standard as employers "with a work place where chemical(s) are produced for use or distribution." An importer is "the first business with employees within the Customs Territory of the United States which receives hazardous chemicals produced in other countries for the purpose of supplying them to distributors or employers within the United States." A distributor is defined as "a business, other than a chemical manufacturer or importer, which supplies hazardous chemicals to other distributors or to employers." For the purposes of the Standard, an employer "means a person engaged in a business where chemicals are either used, distributed, or produced for use or distribution, including a contractor or subcontractor."

OSHA now requires that chemical manufacturers or importers assess the hazards of the chemicals they produce or import in order to pass this information on to employers. The hazard-related information is transmitted in the form of chemical labels and Material Safety Data Sheets (MSDS forms) prepared by the manufacturers or importers. The Standard enumerates certain requirements for information that must appear on both the MSDS forms and chemical labels. Under the Standard, distributors are also required to transmit the MSDS forms and labels to employers. (MSDS forms are discussed in more detail later in this chapter.)

These requirements have a broad application in the workplace. They apply to all chemicals to which employees may be exposed under normal conditions and in foreseeable emergencies. The Standard applies in a more limited way to laboratories and work operations where employees handle only chemicals that are in sealed containers which are not opened under normal circumstances. In both instances, employees are required to maintain the MSDS forms they have received, and to ensure that incoming chemical containers are labeled and the labels are not removed or destroyed.

Employers must also provide their employees with proper information and training.

OSHA does not require labeling of certain categories of chemicals which are already subject to labeling requirements of other federal laws. These include pesticides, foods, food additives, color additives, drugs, cosmetics, distilled spirits, and consumer products.

Certain other substances are exempted from the Standard because they are already regulated under various federal statutes. These include:

- Hazardous waste as defined by the Solid Waste Disposal Act, as amended by the Resource Conservation and Recovery Act of 1976
- Tobacco or tobacco products
- Wood or wood products
- Articles defined as manufactured items "formed to a specific shape or design during manufacturing; . . . which have 'send use' functions dependent in whole or in part upon its shape or design during end use; and . . . which do not release, or otherwise result in exposure to a hazardous chemical under normal conditions of use,"
- Food, drugs, cosmetics, or alcholic beverages found in retail establishments for sale to consumers
- Food, drugs, or cosmetics intended for personal consumption by employees in the workplace

Construction Workers Exposed to Hazardous Chemicals

A. Standard Industrial Classification Major Group No. 15: Building Construction–General Contractors and Operative Builders

Total Employment	Percent of Workers Exposed to Hazardous Chemicals	Number of Exposed Employees
1,137,853	70%	796,497

B. Standard Industrial Classification Major Group No. 16: Construction Other Than Building Construction – General Contractors

Total Employment	Percent of Workers Exposed to Hazardous Chemicals	Number of Exposed Employees
791,882	70%	554,317

C. Standard Industrial Classification Major Group No. 17: Construction – Special Trade Contractors

Total Employment	Percent of Workers Exposed to Hazardous Chemicals	Number of Exposed Employees
2,406,916	70%	1,684,841

Figure 2.1

- Consumer products or hazardous substances as defined in the Consumer Products Safety Act and the Federal Hazardous Substances Act for consumer use in the workplace
- Drugs that can be used by consumers as defined in the Federal Food, Drug, and Cosmetic Act

Hazard Determination

Chemical manufacturers or importers are required to assess the hazards of the chemicals they produce or import, and to pass along such information to subsequent users. The Standard does not require employers to conduct this same assessment. If, however, an employer chooses to disregard the manufacturer's or importer's chemical evaluation, then the employer is required to prepare its own assessment.

For the purpose of identifying health hazards, evidence which is "statistically significant and which is based on at least one positive study conducted in accordance with established scientific principles" is considered sufficient if the results of the study meet the definitions of the "health hazards," as defined by the Standard. The term "health hazard" includes chemicals which are carcinogens, toxic agents, reproductive toxins, irritants, corrosives, sensitizers, hepatotoxins, nephrotoxins, neurotoxins, agents which act on the hematopoietic system, and agents that damage the lungs, skin, eyes, or mucous membranes. Figure 2.2 defines each category of hazard.

The appendices to the Standard provide further guidance for evaluating chemicals. Appendix A to the Standard defines and explains health hazards. Appendix B describes the criteria that must be used to determine whether or not a chemical is to be considered "hazardous." Appendix C lists available data sources recommended by OSHA for use in evaluating potentially hazardous chemicals.

Labels and Other Forms of Warning

One critical aspect of the Hazard Communication Standard is the labeling requirement. Labeling a container of hazardous chemicals can be one of the most direct means of putting an employee on notice that he or she may be exposed to an actual or potential hazard.

Manufacturers', Importers', and Distributors' Requirements

The Standard requires that chemical manufacturers, importers, or distributors ensure that each container of hazardous chemicals leaving the workplace is labeled, tagged or marked. The label must include the following information:

- The identity of the hazardous chemical(s)
- An appropriate hazard warning
- The name and address of the chemical manufacturer, importer, or other responsible parties. (Responsible parties are those persons "who can provide additional information on the hazardous chemical and appropriate emergency procedures, if needed.")

The manufacturers, importers, or distributors must ensure that the labeling is not in conflict with the requirements of the Hazardous Materials Transportation Act and the regulations promulgated under that act. Moreover, if the chemical to be labeled is regulated by OSHA, and is subject to what is known as a "substance-specific health standard," the labeling must also comply with the requirements of that standard. Finally, manufacturers, importers, or distributors are not required to attach new labels to chemical containers if the labels already affixed to the containers provide the required information.

Employer Requirements

Employers are prohibited under the Standard from removing or defacing labels that are affixed to incoming containers, unless such employers immediately re-mark the containers with the required information. The employer must also make certain that the labels are legible, in English, and prominently displayed on the container. As an alternative to actually affixing labels to individual stationary process containers, employers are permitted to

Definitions of Hazards	
1. CARCINOGEN	A cancer-causing substance
2. CORROSIVE	A material that destroys human tissue or metals upon contact
3. HEMATOPOIETIC SYSTEM	A system in the human body that produces or carries blood or blood constituents
4. HEPATOTOXIN	A material that adversely affects the function of the liver
5. IRRITANT	A material that causes discomfort such as tearing, choking, vomiting, skin rashes, or itching
6. NEPHROTOXIN	A material that adversely affects the function of the kidneys
7. NEUROTOXIN	A material that adversely affects the function of the nervous system or brain
8. REPRODUCTIVE TOXIN (TERATOGEN)	A material that adversely affects any of the human reproductive system organs or causes malformation of the fetus
9. SENSITIZER	A substance that causes a person to become prone to an adverse health effect from a different substance, even though the sensitizer itself may provoke no outwardly adverse effect
10. TOXIC AGENT	A substance that produces an adverse health effect on humans

Figure 2.2

use signs, placards, process sheets, batch tickets, operating procedures, or other written material to identify the applicable container and convey the required information. Furthermore, employers are not required to prepare new labels if incoming containers have labels affixed to them which comply with the informational requirements contained in the Standard.

There are two exceptions to the labeling requirements found in the Hazard Communication Standard. First, employers are not required to label portable containers which are used for transferring hazardous chemicals from labeled containers for the immediate use of the employee who performs the transfer. Second, labeling requirements are relaxed with respect to solid metals, such as steel beams or metal castings. While labels are required for initial shipment of the metal to the customer, they are not required for subsequent shipments to the same employer, so long as the information pertaining to the metal does not change.

Material Safety Data Sheets

Chemical manufacturers and importers must obtain or develop a Material Safety Data Sheet for each hazardous chemical they produce or import. Employers are required to have an MSDS for each hazardous chemical they use. An MSDS is required to be in English and to contain at least the following information:

- The identity used on the product label including:
  - The chemical and common name(s) of chemicals which are single substances
  - If the chemical is a mixture which as a whole is hazardous, the chemical and common name(s) of the ingredients that contribute to the hazards, and common name(s) of the mixture
  - If the chemical is a mixture which has not been tested as whole, the MSDS must list:
    - The chemical and common name(s) of all ingredients posing health hazards which comprise 1% or more of the composition, and carcinogenic chemicals making up 0.1% or greater of the mixture
    - The chemical and common name(s) of all hazardous ingredients that comprise less than 1% of the mixture – if the ingredient can be released from the mixture in concentrations exceeding OSHA's permissible exposure limit or the ACGIH (American Conference of Governmental Industrial Hygienists) Threshold Limit Value, or could present a health hazard
    - The chemical and common name(s) of ingredients in the mixture which pose physical hazards
- Physical and chemical characteristics of the chemical (i.e., vapor pressure, flash point)
- Physical hazards of the chemical (i.e., potential for fire, explosion, reactivity)
- Health hazards of the chemical (i.e., signs and symptoms of exposure, medical conditions aggravated by exposure)
- The primary route(s) of entry
- The OSHA permissible exposure limit, ACGIH Threshold Limit Value, and other available exposure limits
- Whether the chemical is listed in the National Toxicology Program Annual Report of Carcinogens (latest edition) or found to be a potential carcinogen in the International Agency for Research on Cancer Monographs (latest edition), or by OSHA
- All precautions for handling and using the chemical (i.e., appropriate

hygienic practices, protective measures for repair and maintenance of contaminated equipment, procedures for cleanup of spills and leaks)

- All applicable control measures (i.e., appropriate engineering controls, work practices, personal protective equipment)
- Emergency and first-aid procedures
- The date of preparation of the MSDS or the last change to it
- The name, address, and telephone number of the manufacturer, importer, employer, or other responsible party preparing or distributing the MSDS who can provide additional information on the chemical if necessary

Note: If the manufacturer, importer, or employer preparing the MSDS cannot obtain any of the items of information listed above, this should be indicated on the MSDS.

Figure 2.3 shows a standard MSDS form. Other variations of MSDS forms are also acceptable. In addition to providing this extensive list of data, the drafter of an MSDS must ensure that the information accurately reflects scientific evidence used in making a hazard determination. Moreover, if at any point after preparing an initial MSDS, new information regarding the chemical's hazards becomes available, this data must be added to the MSDS and distributed to subsequent users within three months.

Manufacturers or importers must provide distributors and employers with the appropriate MSDS forms either prior to, or at the time of, the initial shipment of a chemical. The latest version of the MSDS should also accompany each additional shipment.

The Standard permits the drafters of MSDS forms to choose the format of the MSDS document. The Standard provides that MSDS documents may cover groups of hazardous chemicals in a work area where it may be more appropriate to address the hazards of a "*process* rather than individual hazardous chemicals." Where complex mixtures have similar hazards and contents, one MSDS may be prepared which applies to all of the similar mixtures.

Regardless of the form chosen for the MSDS, the Hazard Communication Standard emphasizes the importance of the employer's obligation to "ensure that in all cases the required information is provided for each hazardous chemical and is readily accessible during each work shift to employees when they are in their work area(s)." Thus, once employers receive MSDS forms, they must keep copies for each hazardous chemical in the workplace, readily accessible to their employees. If the nature of the job is such that employees travel between different workplaces during a work shift, the MSDS forms may be kept at a central location at the employer's primary facility.

Finally, the Standard requires that MSDS forms be made available upon request to designated representatives of employees, the Director of the National Institute for Occupational Safety and Health, the U.S. Department of Health and Human Services (the "Director"), and to the Assistant Secretary of Labor for Occupational Safety and Health, U.S. Department of Labor (the "Assistant Secretary").

Written Hazard Communication Program

The Standard dictates that employers must prepare a Written Communication Program. Essentially, this document must describe how the employer will meet the labeling, MSDS, and employee information and training requirements. In addition, it must include a complete list of the hazardous chemicals present at the work site. This list must identify those chemicals in the same manner as they are referred to in the MSDS. The

Material Safety Data Sheet
May be used to comply with
OSHA's Hazard Communication Standard,
29 CFR 1910.1200. Standard must be
consulted for specific requirements.

U.S. Department of Labor
Occupational Safety and Health Administration
(Non-Mandatory Form)
Form Approved
OMB No. 1218-0072

IDENTITY *(As Used on Label and List)*	*Note: Blank spaces are not permitted. If any item is not applicable, or no information is available, the space must be marked to indicate that.*

Section I

Manufacturer's Name	Emergency Telephone Number
Address *(Number, Street, City, State, and ZIP Code)*	Telephone Number for Information
	Date Prepared
	Signature of Preparer *(optional)*

Section II — Hazardous Ingredients/Identity Information

Hazardous Components (Specific Chemical Identity; Common Name(s))	OSHA PEL	ACGIH TLV	Other Limits Recommended	% *(optional)*

Section III — Physical/Chemical Characteristics

Boiling Point		Specific Gravity (H_2O = 1)	
Vapor Pressure (mm Hg.)		Melting Point	
Vapor Density (AIR = 1)		Evaporation Rate (Butyl Acetate = 1)	
Solubility in Water			
Appearance and Odor			

Section IV — Fire and Explosion Hazard Data

Flash Point (Method Used)	Flammable Limits	LEL	UEL
Extinguishing Media			
Special Fire Fighting Procedures			
Unusual Fire and Explosion Hazards			

(Reproduce locally) OSHA 174, Sept. 1985

Figure 2.3

Section V — Reactivity Data

Stability	Unstable		Conditions to Avoid
	Stable		

Incompatibility (*Materials to Avoid*)

Hazardous Decomposition or Byproducts

Hazardous Polymerization	May Occur		Conditions to Avoid
	Will Not Occur		

Section VI — Health Hazard Data

Route(s) of Entry:	Inhalation?	Skin?	Ingestion?

Health Hazards *(Acute and Chronic)*

Carcinogenicity:	NTP?	IARC Monographs?	OSHA Regulated?

Signs and Symptoms of Exposure

Medical Conditions
Generally Aggravated by Exposure

Emergency and First Aid Procedures

Section VII — Precautions for Safe Handling and Use

Steps to Be Taken in Case Material Is Released or Spilled

Waste Disposal Method

Precautions to Be Taken in Handling and Storing

Other Precautions

Section VIII — Control Measures

Respiratory Protection *(Specify Type)*

Ventilation	Local Exhaust	Special
	Mechanical *(General)*	Other

Protective Gloves	Eye Protection

Other Protective Clothing or Equipment

Work/Hygienic Practices

Page 2

☆ U.S.G.P.O.: 1986-491-529/45775

Figure 2.3 (cont.)

Program must also set forth the methods the employer intends to use to inform employees of hazards associated with non-routine tasks and chemicals contained in unlabeled pipes.

Of particular importance to the construction industry is the further requirement that employers involved with contractors, their employees, subcontractors, and their employees at particular work sites, identify the methods the employer intends to use to make MSDS's available to the other employers. The Program must also identify the methods the employer will use to inform the other employers of precautionary methods that might be necessary and the methods used to inform the other employers of the labeling system utilized at the workplace. If a Written Hazard Communication Program already exists, an employer may rely on it rather than preparing a new one, as long as the existing program complies with the criteria of the Hazard Communication Standard. Chapter 3 describes the requirements of a Hazard Communication Program in detail. As with the MSDS form, the Hazard Communication Program must be available upon request to employees, their designated representatives, the Assistant Secretary, and the director.

Employee Information and Training

The Standard requires that employers provide their employees with information and training on the hazardous chemicals in the workplace–at the time of their initial assignment and whenever new hazards are identified. Employers must inform their employees of all the requirements of the Hazard Communication Standard, all operations in their work area where hazardous chemicals are present, and the location and availability of MSDS forms and the Written Hazard Communication Program.

The training requirements contained in the Hazard Communication Standard are quite specific. Training programs for employees beginning their initial assignments and for those faced with new hazards must include instructions on the methods used to detect the presence or release of hazardous chemicals in the workplace (i.e., monitoring conducted by the employer, monitoring devices, visual appearance of hazardous chemicals).

Employees must also be instructed as to the physical and health hazards of the chemicals, and informed of the measures that can be taken to protect themselves from such hazards. Such measures may include appropriate work practices, emergency procedures, and protective equipment.

Furthermore, employees must be informed of the details of the written Hazard Communication Program prepared by the employer, and the usefulness and accessibility of the Program. Chapter 3 describes both the Hazard Communication Program and the hazard communication training program in more detail.

Trade Secrets

Manufacturers, importers, or employers are permitted, under certain circumstances, to exclude the chemical name and other identifying descriptions of hazardous chemicals from the information they are required to provide. They may withhold a chemical's identity only if divulging it would result in the disclosure of a trade secret. The manufacturer, importer, or employer wishing to withhold a chemical's identity may not do so unless his or her claim that the withheld information is a trade secret can be supported by specific evidence. One must be able to demonstrate that the withheld information involves a confidential formula, pattern, process, device, information, or compilation of information that gives the employer an advantage over competitors who do not know or use it.

In situations where manufacturers, importers, or employers have withheld the specific chemical identity of a substance, they are obligated to indicate the following information on the MSDS:

- That the chemical's identity is being withheld as a trade secret
- The properties and effects of the chemical
- All other precautionary information required under the standard.

The right to withhold a chemical's identity is not unconditional. Under certain circumstances, the chemical identify *must* be made known. For example, in emergency situations where an exposed individual must be treated, the identity of the chemical must be revealed upon request by the treating physician or nurse. Even in non-emergency situations, health professionals providing health services to employees who have been exposed to chemicals can request that the chemical identity be disclosed. Such a request must be in writing and must describe the specific reasons why disclosure is essential. Justifiable grounds for disclosure of the identities of trade secret chemicals in non-emergency situations include the following.

- To assess the hazards associated with the chemical
- To conduct samplings of the workplace atmosphere in order to determine employee exposure level
- To conduct medical surveillance of exposed employees
- To select appropriate protective equipment for exposed employees
- To design protective measures for exposed employees
- To conduct studies of health effects on exposure

Moreover, the health professional, employee, or a designated representative must also demonstrate that the disclosure of the properties and effects of the chemical, measures for controlling exposure, monitoring and analyzing exposure, diagnosing and treating exposure without disclosing the chemical's identity would not be sufficient.

In both emergency and non-emergency situations, manufacturers, importers, or employers may request that a written *Statement of Need* and a *Confidential Agreement* be prepared and signed by the involved parties. Under the Standard, the Confidential Agreement may restrict the use of the information to health purposes and may provide for appropriate legal remedies in the event of a breach of the Agreement. Finally, in situations where the manufacturer, importer, or employer has denied a request for disclosure, the requesting party may refer the denial to OHSA. OSHA will then determine whether disclosure is necessary and take further action as appropriate.

Chapter 3

Hazard Communication Program

The Federal Occupational Safety and Health Act (OSHA) was amended on August 24, 1987 to apply the Hazard Communication Standard to non-manufacturing employers, including construction companies. Under the standard, employers who employ ten or more persons are required to establish a written hazard communication program. In addition, employers must maintain written information about the hazardous substances on their work site and transmit that information to their employees. The means of information transfer are specified to include labels on containers, material safety data sheets (MSDS forms), and employee training programs.

The Hazard Communication Standard is described in detail in Chapter 2. This chapter will describe the hazard communication program prescribed by the standard. The first section will identify the responsibilities of the various individuals involved. The second section will outline the required hazard communication program elements. The third section will outline the required training program elements and suggest ways to effectively implement the program.

The Hazard Communication Standard imposes specific responsibilities for employee safety. The responsibilities are separated in the Hazard Communication Program into employer and employee responsibilities. It is important to note that the employee does bear some responsibility for his/her own safety. The employer must provide the information, the tools, the incentive, and the opportunity; but the employee must take the final protective action.

Employer Responsibilities

It is the employer's responsibility under the Hazard Communication Standard to provide the employee with the knowledge and means to protect his/her own safety. The Standard specifies the following particular employer responsibilities.

- Employers must prepare a complete inventory of all hazardous chemicals and materials on the job site, and then list all the potential chemicals and materials that will be used on that site. This inventory must be updated on a regular basis to reflect changes as new products are received and products no longer needed are removed from the job site. The inventory must be kept in a central location on the job site, with individual lists of specific chemicals and materials in appropriate locations within the various work areas. Both the master inventory and the individual work area lists must be accessible at all times to employees who are working with those chemicals or materials.

- Employers must ensure that the MSDS forms are collected for all the hazardous chemicals and materials on the job site. They must also keep a central file of all MSDS forms on site. This file is to be placed in locations that are readily accessible to the employees during each work shift. It is recommended that both the inventories of hazardous chemicals and materials and MSDS forms, be kept together at the central and accessible locations.
- Employers must provide an MSDS form to any employee, or to the employee's designated representative, for any material to which that employee is exposed. This must be done within fifteen (15) days of receipt by the employer of a written request for the form.
- Employers are responsible for labeling all containers of hazardous chemicals and materials on the work site. Any container of hazardous material that arrives at the job site without proper labeling must be properly labeled by the employer upon its receipt.
- Employers are responsible for providing information and training on the safe handling and proper use of hazardous chemicals and materials in the employee's work area at the time of his/her initial assignment, and whenever a new hazardous chemical and material is brought onto the specific site or work area.
- Employers who are using trade secret chemicals subject to trade secret protection, as defined by the law, must still provide specific health risk information on the trade-secret chemical to an employee's treating physician or nurse upon written request. In an emergency situation, this request may be oral and must be complied with immediately.

Employee Responsibilities

A responsible employer will provide the necessary information to his/her employees to allow them to protect their own health and safety on the job site. He or she will also provide the necessary safety tools and equipment, establish policies requiring use of the procedures and equipment provided, and establish incentives or penalties to encourage employee participation. The employee, however, bears ultimate responsibility for personal safety. The employee must use the information and tools provided by the employer. Specifically, the Standard imposes the following responsibilities on the employee.

- Employees must actively participate in the training programs and comply with training provisions.
- Employees must become familiar with the material safety data sheets and labeling systems of those hazardous materials with which they work.
- Employees must make use of protective control measures identified to them and protect themselves from adverse exposure to hazardous materials.
- Employees must notify supervisors of the presence of unlabeled containers.
- Employees must notify supervisors of any MSDS's which are missing from the work area file.

The responsibilities of both employer and employee are summarized in Figure 3.1.

Purpose and Contents of the Hazard Communication Program

The Hazard Communication Standard issued by OSHA is intended to establish minimum criteria upon which to base the preparation of individual Hazard Communication Programs. The programs are site-specific. Promulgation of standard requirements, which all programs must meet, ensures that all employees will receive at least the minimum information necessary to protect their health and safety.

Program contents are defined by the OSHA Hazard Communication Standard. The Standard, described in complete detail in Chapter 2, can be found in the Code of Federal Regulations, Chapter 29, Sections 1910, 1915, 1917, 1918, 1926, and 1928. The requirements for the program, as specified by the Standard, are listed below. Figure 3.2 is a summarized version.

1. Employers must develop and implement a written hazard communication program for the workplace, which includes at least the following items.
   a. A list of the hazardous chemicals known to be present. Each chemical should be identified using the same terminology as used on the appropriate Material Safety Data Sheet. The list may be compiled for the workplace as a whole or for individual work areas.
   b. The methods the employer will use to inform employees of the hazards of non-routine tasks (such as cleaning reactor vessels), and the hazards associated with chemicals contained in unlabeled pipes in their work areas.

Responsibilities under the Hazardous Communication Program

Employer Responsibilities

1. Conduct inventory of site
2. Collect MSDS forms
3. Make MSDS forms available to employees
4. Label all containers
5. Conduct new employee health/safety training
6. Provide health information on trade secret materials in emergency
7. Provide initial health/safety training to all employees
8. Provide safe work environment

Employee Responsibilities

1. Participate in training
2. Learn what MSDS forms are and how to use them
3. Make use of safety equipment provided
4. Notify supervisor about unlabeled containers
5. Notify supervisor about missing MSDS forms

Figure 3.1

c. The methods the employer will use to inform any contract employers whose employees will be working in the employer's workplace. These contract employees must be informed of the hazardous chemicals to which they will be exposed while performing their work. They should also be provided with recommendations for appropriate protective measures.

2. The employer may rely on an existing hazard communication program to comply with these requirements, provided it meets the criteria established above.
3. The employer must make the written hazard communication program available, upon request, to the employees, their designated representatives, the Assistant Secretary of OSHA, and the Director of OSHA.
4. A description of how the following requirements will be met.
   a. Except as provided in paragraphs b and c below, the employer must ensure that each container of hazardous chemicals in the workplace is labeled, tagged, or marked with the following information:
      (i) The identity of the hazardous chemical(s) contained therein
      (ii) Appropriate hazard warnings
   b. The employer may use signs, placards, process sheets, batch tickets, operating procedures, or other such written materials in lieu of affixing labels to individual stationary process containers, as long as the alternative method identifies the containers to which these documents apply, and conveys the information required by paragraph 4a above. The written materials must be readily accessible to the employees in their work area throughout each work shift.

Hazard Communication Program

1. Must be written, and must include:
   a. Inventory of hazardous materials in workplace
   b. Methods to be used to inform employees of risks
   c. Methods to be used to inform contractors on site
2. Must be available to all employees
3. Must explain how the following will be met:
   a. All containers to be labeled
   b. Existing labels not to be defaced
   c Labels to be in English, but may also be in any other appropriate language
   d. MSDS forms to be obtained and maintained
   e. Access to MSDS forms to be provided for all employees
   f. Employee Right-to-Know training to be provided

Figure 3.2

c. The employer is not required to label portable containers into which hazardous chemicals are transferred from labeled containers, if such portable containers are intended only for the immediate use of the employee who performs the transfer.

d. The employer must not remove or deface existing labels on incoming containers of hazardous chemicals, unless the container is immediately marked with the required information.

e. The employer must ensure that labels or other forms of warning are legible and in English. Such warnings must be available in the work area throughout each work shift. Employers whose employees speak other languages may add the warnings in these other language(s), as long as the information is presented in English as well.

f. The employer need not affix new labels to comply with this section if existing labels already convey the required information.

g. Employers must obtain and maintain a Material Safety Data Sheet for each hazardous chemical used by the employees.

h. The employer must maintain copies of the required Material Safety Data Sheets for each hazardous chemical in the workplace, and must ensure that they are readily accessible during each work shift to employees when they are in their work area(s).

i. Material Safety Data Sheets (MSDS) may be kept in any form (for example, they may be included with standard operating procedures). An MSDS may be designed to cover *groups* of hazardous chemicals in a work area where it may be more appropriate to address the hazards of a process, rather than individual hazardous chemicals. However, the employer must ensure that in all cases, the required information is provided for each hazardous chemical, and is readily accessible during each work shift to employees in their work area(s).

j. Employers must provide employees with information and training on hazardous chemicals in their work area at the time of their initial assignment, and whenever a new hazard is introduced into their work area.

   (i.) Employees must be informed of:

      (a) The requirements of the hazard communication program

      (b) Any operations in their work area where hazardous chemicals are present

      (c) The location and availability of the written hazard communication program, including the required list(s) of hazardous chemicals, and Material Safety Data Sheets

   (ii.) Employee training must include at least:

      (a) Methods and observations that may be used to detect the presence or release of a hazardous chemical in the work area (such as monitoring conducted by the employer, continuous monitoring devices, visual appearance or odor of hazardous chemicals when being released, etc.)

      (b) Physical and health hazards of the chemicals in the work area.

(c) Measures employees can take to protect themselves from these hazards, including specific procedures the employer has implemented to protect employees from exposure to hazardous chemicals, such as appropriate work practices, emergency procedures, and personal pretective equipment to be used
(d) Details of the hazard communication program developed by the employer, including an explanation of the labeling system and the Material Safety Data Sheet, and how employees can obtain and use the appropriate hazard information

Hazard Communication Training

An appropriate employee training program must be presented to each employee at the time he or she is hired, whenever the employee is assigned to a new job or work area, and whenever a new hazard is introduced into the work area. The training program should include the following elements, which are summarized in Figure 3.3.

- An introduction explaining the purpose of the training
- The requirements of the Hazard Communication Standard
- The location and manner of access to the written hazard communication program, which includes the inventory of hazardous chemicals and the Material Safety Data Sheets
- Operations or processes in the work environment where hazardous chemicals or materials are present
- Physical and health hazards of chemicals and materials in the work area
- Methods and observations that can be used to detect the presence or release of a hazardous chemical or material in the work area
- Measures employees can take to protect themselves
- An explanation of the labeling system being used
- An explanation of Material Safety Data Sheets
- Methods to extract and utilize appropriate hazard information from labels, MSDS forms, and other pertinent sources

Hazard Communication Training Program Outline

1. Introduction
2. Requirements of the Hazard Communication Standard
3. Location and access to Inventory and MSDS sheets
4. Identification of hazardous activities in the workplace
5. Health hazards from materials in the workplace
6. Methods for detecting exposure to hazards
7. Methods for protecting against exposure
8. Explanation of container labeling system
9. Explanation of MSDS forms
10. How to read an MSDS and how to use information

Figure 3.3

Methods

Training programs can usually be standardized through the use of common teaching tools. Several films have been produced and are commercially available, which can provide an introduction to the training session and the requirements of the Hazard Communication Standard. A brief lecture is best used to describe the location and manner of access to the program on the job site and to identify the operations or processes in the work area where hazardous chemicals or materials will be present. A second film can be used to describe the hazards associated with various types of chemicals and materials, the methods which can be used to detect the release of a hazardous substance, and the personal protective measures that are available. Labeling is usually standardized, so that a film on that subject, supported by examples from the work site, and followed by a film on the use of MSDS forms, will round out a training program.

Clearly, only the site-specific information regarding the hazards to be found on a particular work site needs to be addressed on an individual basis. Even this aspect can be covered with a training film, provided conditions are reasonably consistent from site to site.

Training should be conducted in a well-lighted, well-ventilated area, free of excessive noise or distractions. The entire session should last no more than one to one-and-a-half hours. Comfortable seats, set far enough apart to avoid crowding adds to the attention span of the participants, and coffee or other refreshments will encourage active participation by employees.

One final note on training sessions: if the employee senses that management is conducting a training session merely to keep OSHA happy, the employee will not feel compelled to participate in program implementation. However, when management takes a strong, active, and aggressive role in program implementation, employees sense the concern for their health and safety inherent in the action and tend to pay more attention to what is being taught. By taking the time to develop an effective training manual which can be given to each employee during the training session, and by actively participating in the program, management can better promote its program policies.

Chapter 4

Storage of Hazardous Materials

Effective management of hazardous **materials** is key to the effective management of hazardous **wastes.** Management of hazardous materials begins with the proper storage and control of incoming inventories. This chapter provides detailed descriptions of how to safely store the various types of hazardous materials used on a construction site. Schematic layouts of typical storage area facilities are provided, along with detailed construction drawings for storage bins, storage racks, and fire-resistant enclosures. Also included are a bill of materials for each illustrated storage unit.

Basic Principles

The basic principles to follow when storing hazardous materials are:

- Protection from physical damage to the containers
- Isolation of materials from other, incompatible materials
- Isolation of flammables from heat, flames and sparks

The most difficult principle to put into practice is isolating incompatible materials. The problem is that most people know very little about the compatibility of substances and materials.

Three approaches can be taken to overcome this problem.

- First, and most important, read the Material Safety Data Sheet (MSDS) provided by the manufacturer for every material shipped to the site. MSDS *forms are required to list a complete description of all incompatible materials.*
- Second, follow the rule: "when in doubt, isolate."
- Third, review Figures 4.1 and 4.2, prepared by the Defense Logistics Agency of the Department of Defense. Figure 4.1 identifies materials based on compatibility groups and assigns a code to each group. Figure 4.2 shows a recommended hierarchy of storage groups. Again, the MSDS form will help identify the category or group into which a given material falls. After identifying the appropriate compatibility code from Figure 4.1, turn to Figure 4.2. Figure 4.2 shows recommended storage groupings by compatibility code. Compatibility groups shown in the same box can be safely stored together. Those in separate boxes must be stored in separate areas. Note that it is not necessary to store compatible groups together, but that it is safe to do so.

It should also be noted that when the Defense Logistics Agency indicates *storage within the same area* (see Figure 4.3), this means that materials in different compatibility groups should be separated by at least four feet, except for those materials in compatibility code groups L1, M1, and N1; which can be stored more closely together. While not defined by the Defense Logistics Agency, *separate storage areas* can be construed to mean separation by at least ten feet or, better yet, by a physical barrier or wall.

Compatibility Codes for Hazardous Material Storage

Compatibility Groups	Codes	Compatibility Groups	Codes
1. Flammable/Combustible Liquids		5. Compressed Gases	
a. Class Ia	F1	a. Flammable–Toxic or Poison A	G1
b. Class Ib	F2	b. Flammable–Non Toxic	G2
c. Class Ic	F3	c. Non Flammable–Toxic or Poison A	G3
d. Class II	F4	d. Non Flammable–Non Toxic	G4
e. Class IIIa	F5	e. Chlorine	G5
f. Class IIIb	F6	f. Oxygen or Oxidizers	G6
2. Corrosives		g. Acetylene	G7
a. Acids			
(1) Organic	C1	6. Radioactive Materials/Devices	A1
(2) Inorganic	C2		
b. Bases		7. Explosives	
(1) Organic	B1	a. Class A	E1
(2) Inorganic	B2	b. Class B	E2
		c. Class C	E3
3. Reactive Chemicals			
a. Oxidizers	R1	8. Irritants	
b. Reducers	R2		
c. Water Reactive	R3	9. Low Hazard	L1
d. Pyrophoric (Flammable Solids)	R4		
		10. Magnetic Material	M1
4. Toxic Chemicals			
a. Pesticides		11. Not Hazardous	N1
(1) Herbicides	P1		
(2) All Others (Insecticides, etc.)	P2	12. Special Chemicals or Multiple Hazardous	S1
b. Non-Pesticides			
(1) Carcinogens	T1		
(2) Bioaccumulatives	T2		
(3) Acute Toxic Chemicals	T3		
(4) Chronic Toxic Chemicals	T4		
(5) Etiologic Agents	T5		

NOTE: *Local segregation requirements apply.*
From: "Hazardous Materials Storage and Handling Handbook," Defense Logistics Agency, Dept. of Defense, Cameron Station, Alexandria, VA, 27 April 1984, pp. 43-44.

Figure 4.1

This chapter will define and discuss the following substances. The uses, physical hazards, and storage requirements for these substances are all covered.

- Acids
- Alkalis
- Cholorinated solvents
- Chlorine
- Compressed gases
- Coolants
- Flammable solvents
- Fuel (gasoline/diesel)
- Paints and thinners
- Phenolic compounds
- Petroleum products (oils/lubricants)

Acids

Acids are most commonly used on a construction site as cleaners for brickwork and masonry and as an electrolyte in the batteries of heavy equipment and vehicles. Types of acid used on the site might include muriatic acid, which is a common name for a weak solution of hydrochloric acid, and sulphuric acid, which is the base ingredient in battery acid.

Hierarchy of Storage Groups

Groups	Codes	Groups	Codes
1. Radioactive Material	A	20. Pyrophoric	R4
2. Explosive Class A	El	21. Water Reactive	R3
3. Explosive Class B	E2	22. Reducers	R2
4. Explosive Class C	E3	23. Organic Acid	Cl
5. Etiologic Agents	T5	24. Organic Base	Bl
6. Special Chemicals/Multiple Hazardous	S	25. Inorganic Acid	C2
7. Herbicides	Pl	26. Inorganic Base	B2
8. All Other Pesticides	P2	27. Carcinogens	Tl
9. Flammable–Toxic Poison A Gas	Gl	28. Bioaccumulatives	T2
10. Non Flammable–Toxic/Poison A Gas	G3	29. Acute Toxic Chemicals	T3
11. Flammable–Non Toxic Gas	G2	30. Chronic Toxic Chemicals	T4
12. Acetylene	G7	31. Irritants	J
13. Oxygen/Oxidizers	G6	32. Class II Combustible Liquid	F4
14. Chlorine	G5	33. Class IIIa Combustible Liquid	F5
15. Non Flammable–Non Toxic Gas	G4	34. Class IIIb Combustible Liquid	F6
16. Class Ia Flammable Liquid	Fl	35. Low Hazard	L
17. Class Ib Flammable Liquid	F2	36. Magnetic Materials	M
18. Class Ic Flammable Liquid	F3	37. Not Hazardous	N
19. Oxidizers	Rl		

NOTE: *Group hierarchies are used to determine the placement of items in storage when more than one compatibility code applies.*
From: *"Hazardous Materials Storage and Handling Handbook," Defense Logistics Agency, Dept of Defense, Cameron Station, Alexandria, VA,* 27 April 1984, *pp.* 45-46.

Figure 4.2

Health Hazards

Skin, eyes, and clothing may become burned when they come into direct contact with liquid acid. Acid fumes, if inhaled, may burn the respiratory tract, including the mouth, nose, throat, and lungs. If acids are involved in a fire, they can release fumes that are even more toxic than the acid itself. Lungs burned with acid fumes often develop pulmonary edema, a condition in which the lungs fill with fluid, making breathing extremely difficult. Untreated acid burns, particularly those resulting from ingestion and inhalation, are usually fatal.

Physical Hazards

Acids react violently with alkalis, cyanides, sulfides, chlorinated organics, flammable organics, and metals. They become more aggressive at elevated temperatures and boil explosively if inundated with water. The goal of a safe storage plan for acids is to avoid inadvertent contact with either people or incompatible compounds.

Storage

There are two important concepts in proper acid storage: **isolation** and **protection.** Isolation keeps the acids from contacting incompatibles, and protection prevents damage to containers, thereby avoiding spillage or leakage. The following paragraphs describe suggested practice to ensure isolation and protection.

Chemical Segregation

Recommended Storage by Compatibility Groups

F1, F2, F3 F4, F5, F6	C1, C2 L1, M1, N1	B1, B2 L1, M1, N1	R2, T3, T4 L1, M1, N1
R1, R4, L1, M1, N1	R3	P1, P2	G1, G2, G3 G4, G5
G6	G7	A1	T1, T2
E1, E2, E3	T5	S1	I1

NOTE: *Separate blocks mean separate storage locations. All groups within the same area must be separated by at least 4 ft. except for* L1, M1, N1.

From: "Hazardous Materials Storage and Handling Handbook," Defense Logistics Agency, Dept. of Defense, Cameron Station, Alexandria, VA, 27 April 1984, pp. 47.

Figure 4.3

Small containers of acids, up to five gallons each, should be stored in fire-resistant steel cabinets, such as that shown in Figure 4.4. The cabinet should be stored as far away from other chemicals as is reasonably practicable, but at least ten feet from any incompatible material. A review of Figure 4.3 and the Material Safety Data Sheets supplied by the acid manufacturers will help to identify incompatible materials.

Drums of acid should be stored in an area separate from all other chemicals, or adjacent to a steel cabinet holding smaller containers of acids. If drums are stored in a trailer, it is recommended that a wooden barrier such as that shown in Figure 4.5 be used to protect them from damage caused by movement of other construction materials into and out of the trailer. Note that acids are generally incompatible with aluminum metal. Therefore aluminum studs are not recommended for use in building the protective barrier. The aluminum trailer siding must be protected.

Storage Temperature

All storage areas used for liquid acids must be maintained at a temperature between 35°F and 120°F.

Alkalis

Alkalis are most commonly found on construction sites as bonding agents in cement and mortar. They are also the principal ingredient in industrial strength cleansers and detergents, as well as in drilling fluids and lubricants. Examples include calcium hydroxide, which is a mortar additive; sodium hydroxide, which is a cleanser additive, and calcium oxide, which is a hydrated derivative of calcium hydroxide used in drilling fluids and lubricants. *Calcium hydroxide* is the chemical name of common lime, a soil additive used to enhance lawn growth. Calcium chlorides are often used as dust suppressors and dessicants.

Health Hazards

Alkalis burn the skin when in direct contact, although powder forms are usually not aggressive so long as they remain dry. Dust or liquid alkalis burn the eyes and mucous membranes on contact. Ingestion causes burning of the digestive tract, while inhalation causes burning of the nose, throat, and respiratory tract.

Physical Hazards

Alkalis react violently with acids, particulary when exposed to heat. Dry forms of alkalis tend to be very powdery, which can lead to large dust problems in handling. A safe storage plan for alkalis must account for both dry and liquid forms. The goal is to protect containers from physical damage, a task that may be more difficult for the dry forms since they are usually shipped in paper sacks. Measures must also be taken to keep them from inadvertent contact with people or acids.

Storage

Dry alkalis can be stored on wooden pallets tall enough to prevent accidental wetting from minor flooding. Pallets should be stored not less than three feet apart and not less than ten feet from any acids or sources of water. Pallets and drums should be stored away from active traffic areas in order to avoid accidental damage from passing trucks and building material transporters. A review of the Material Safety Data Sheets and Figure 4.3 will help to identify incompatible materials.

If storing alkalis in a trailer, it is recommended that a wooden barrier, such as that shown in Figure 4.5, be used to protect the materials from damage caused by the movement of construction materials into and out of the trailer. Note that most alkalis are incompatible with aluminum. Therefore, use of

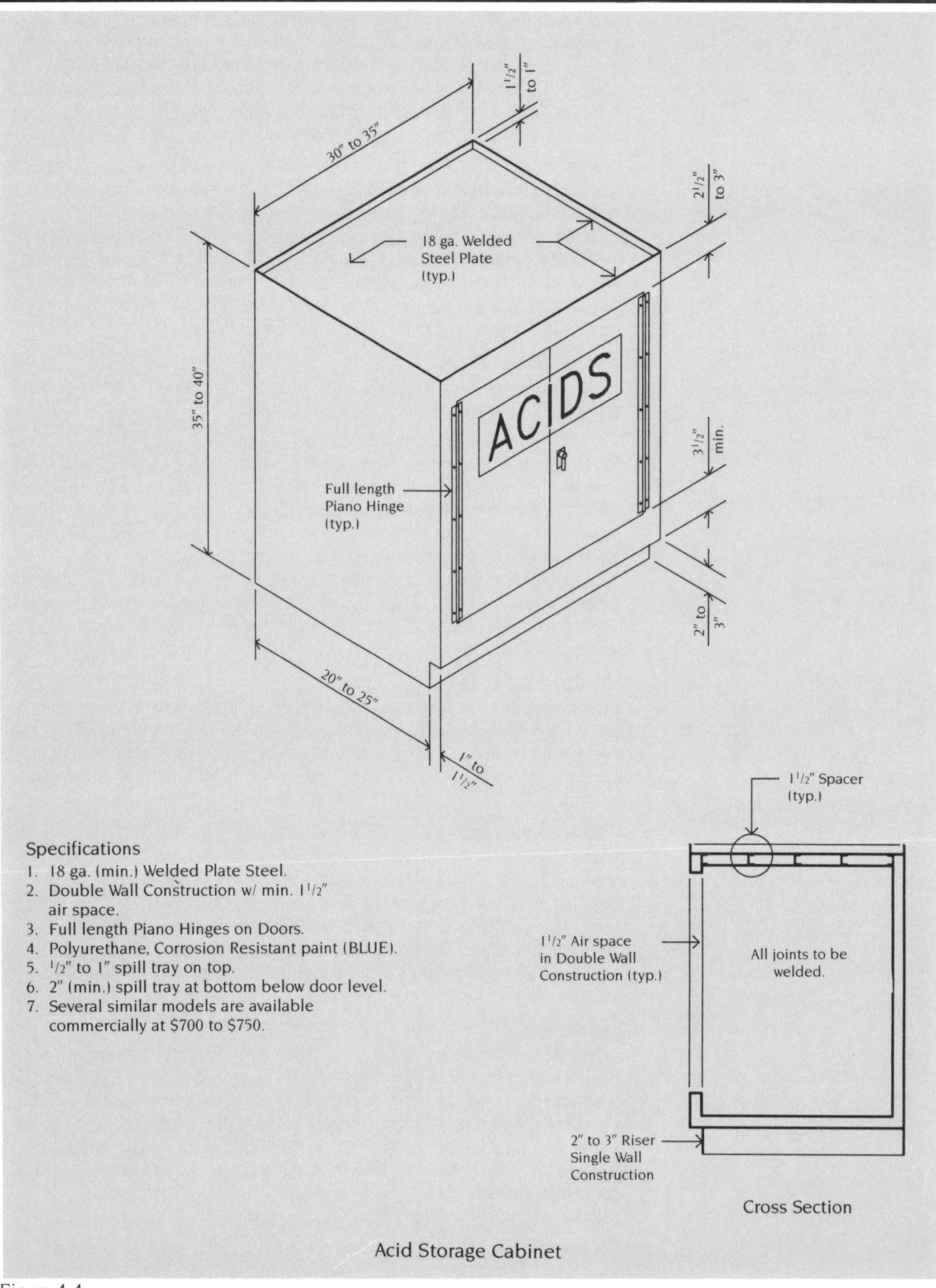

30" to 35"
1/2" to 1"
2 1/2" to 3"
18 ga. Welded Steel Plate (typ.)
35" to 40"
ACIDS
3 1/2" min.
Full length Piano Hinge (typ.)
2" to 3"
20" to 25"
1" to 1 1/2"
1 1/2" Spacer (typ.)
Specifications
1. 18 ga. (min.) Welded Plate Steel.
2. Double Wall Construction w/ min. 1 1/2" air space.
3. Full length Piano Hinges on Doors.
4. Polyurethane, Corrosion Resistant paint (BLUE).
5. 1/2" to 1" spill tray on top.
6. 2" (min.) spill tray at bottom below door level.
7. Several similar models are available commercially at $700 to $750.
1 1/2" Air space in Double Wall Construction (typ.)
All joints to be welded.
2" to 3" Riser Single Wall Construction
Cross Section
Acid Storage Cabinet

Figure 4.4

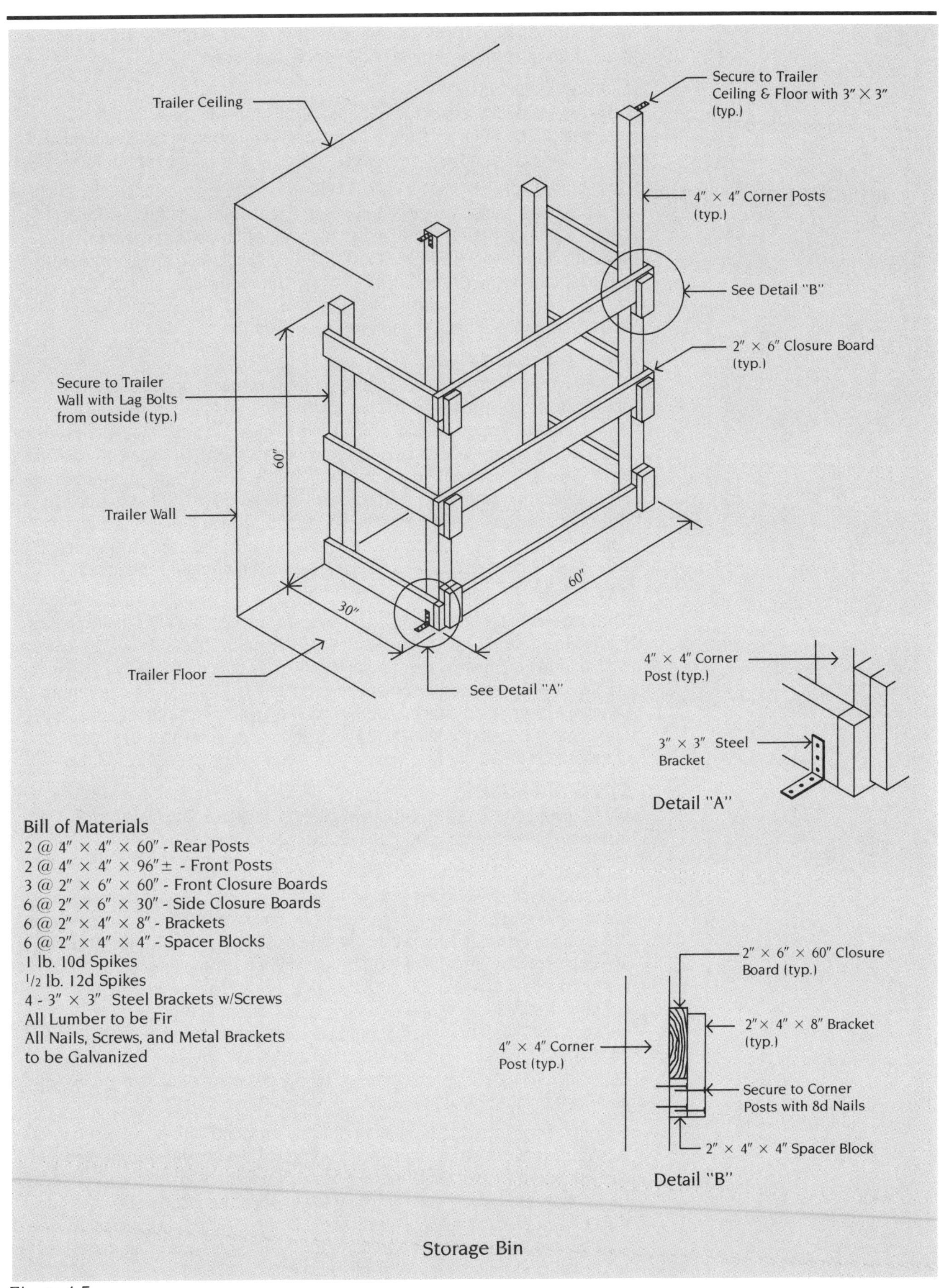
Trailer Ceiling
Secure to Trailer Ceiling & Floor with 3" X 3" (typ.)
4" × 4" Corner Posts (typ.)
See Detail "B"
2" × 6" Closure Board (typ.)
Secure to Trailer Wall with Lag Bolts from outside (typ.)
60"
Trailer Wall
60"
30"
Trailer Floor
See Detail "A"
4" × 4" Corner Post (typ.)
3" × 3" Steel Bracket
Detail "A"
Bill of Materials
2 @ 4" × 4" × 60" - Rear Posts
2 @ 4" × 4" × 96"± - Front Posts
3 @ 2" × 6" × 60" - Front Closure Boards
6 @ 2" × 6" × 30" - Side Closure Boards
6 @ 2" × 4" × 8" - Brackets
6 @ 2" × 4" × 4" - Spacer Blocks
1 lb. 10d Spikes
1/2 lb. 12d Spikes
4 - 3" × 3" Steel Brackets w/Screws
All Lumber to be Fir
All Nails, Screws, and Metal Brackets to be Galvanized
2" × 6" × 60" Closure Board (typ.)
2"× 4" × 8" Bracket (typ.)
4" × 4" Corner Post (typ.)
Secure to Corner Posts with 8d Nails
2" × 4" × 4" Spacer Block
Detail "B"
Storage Bin

Figure 4.5

aluminum studs to construct the protective barrier is not recommended, and protection of aluminum trailer side panels is advised.

Storage Temperature

All areas used for the storage of liquid alkalis must be maintained at a temperature between 35°F and 120°F. Good ventilation is also required.

Chlorinated Solvents

Chlorinated solvents are typically found in on-site maintenance areas, where they are used to clean heavy equipment, engine parts, and tools. These compounds are the principal active ingredient in most degreasers, shop solvents, and machine tool cleaners. Examples include trichloroethylene, tetrachloroethylene (sometimes called perchloroethylene, or "Perc"), trichloroethane (sometimes called "Trichlor"), and methylene chloride, which is a common ingredient in paint removers.

Health Hazards

There are two classes of health effects that may result from exposure to chlorinated solvents: acute and chronic. Acute effects are those which occur very soon after exposure and which tend to clear up soon after the exposure is ended. The acute effects of external contact include chloracne (a special form of common acne); dermatitis; and drying, flaking skin. Ingestion causes rapid poisoning which can lead to death. Inhalation at low levels leads to dizziness, nausea, headaches, and cyanosis (blue skin and lips). At higher concentration, it can lead to coma or death. Most of these compounds volatize at low temperatures, thereby creating a high risk of inhalation exposure for the people working with them.

Chronic effects are those which continue for a long time after the exposure has ended. Chronic effects may occur quickly upon exposure, fail to clear up well, and may represent permanent injury. They include more severe chloracne and dermatitis. Long-term exposure, even at low concentrations, can cause permanent kidney and liver damage. Some chlorinated solvents are suspected carcinogens. There is evidence to suggest that the effects of chlorinated solvents on the body are exacerbated by heavy alcohol use.

Physical Hazards

When involved in a fire, chlorinated solvents generate very toxic fumes including phosgene gas and gaseous hydrochloric acid.

Storage

Safe storage of these materials is characterized by protection from acids, flames and sparks, as well as protection from physical damage to containers. Solvents are generally shipped to the site in 35-gallon or 55-gallon drums, although smaller containers are not uncommon. Drums should be stored on wooden pallets or a solid floor that is not subject to flooding. Smaller containers should be stored in their original shipping cartons on wooden pallets. Containers of solvent should be stored at least ten feet from acids, alkalis, and other incompatible materials. A review of Figure 4.3 and the Material Safety Data Sheets supplied by the solvent manufacturer will help to identify incompatible materials.

Solvent storage areas in a trailer should be provided with a wooden barrier, similar to that shown in Figure 4.5, to protect the containers from damage caused by the movement of other construction materials into and out of the trailer. A solid wooden wall may be used to separate incompatibles inside a trailer when storage space is particularly limited. Note that many chlorinated solvents are incompatible with aluminum. Therefore, use of aluminum studs to construct storage facilities inside a trailer is not recommended and protection of aluminum trailer side panels is recommended.

Storage Temperature

All storage areas for chlorinated solvents must be maintained at a temperature between 35°F and 120°F.

Chlorine

Chlorine is commonly used on construction sites as a disinfectant for new water systems. Chlorine is usually purchased and stored as a hypochlorite crystal, hypochlorite solution or as a compressed chlorine gas. The hypochlorite or chlorine gas is mixed with water just prior to use.

Health Hazards

Chlorine, in any form, reacts rapidly with body moisture to form hydrochloric acid, a substance that burns the adjacent tissue. Hydrochloric acid is extremely irritating to skin, eyes, and mucous membranes. Prolonged exposure may cause chloracne (a special form of common acne). At high concentrations, chlorine acts as an asphixiant, causing choking as a result of swelling of the mucous membranes. Contact with chlorine may also cause nausea, vomiting, and acute respiratory distress.

Physical Hazards

Chlorine gas is highly corrosive. In any damp areas, such as wet or damp floors, steam mists, or exterior storage areas, chlorine will form hydrochloric acids which are also extremely corrosive.

Storage

Dry chlorine crystals can be stored on wooden pallets tall enough to prevent wetting of the crystals from minor flooding. Containers should be protected from precipitation and other water sources. Liquid hypochlorite or chlorine solutions should be stored on wooden pallets in their original shipping containers. Compressed gas cylinders should be stored in a very well ventilated area and protected from physical damage. Chlorine should not be stored near flammable substances, heat, sparks, or open flames, due to the highly toxic gases created during the burning of chlorine compounds. Gas cylinders may be stored with cylinders of other compressed gases except oxygen, oxidizers, and acetylene. Carbon dioxide and other inert gases are compatible with chlorine.

Pallets of hypochlorite should be protected from physical damage due to the corrosive properties of the dry chemical. Cylinders must be protected from physical damage for two reasons: 1) the potential for explosion of the cylinder, and 2) the highly corrosive nature of the free gas.

Storage of small quantities of liquid or crystals in a trailer should be on wooden pallets segregated from acids and oxidizers. A wooden barrier (similar to that shown in Figure 4.5) is recommended to protect the containers from damage. Gas cylinders must be securely lashed to a wall (as shown in Figure 4.6). Note that chlorine is corrosive to most metals. Aluminum studs are not recommended for storage compartment construction, and protection of aluminum trailer panels is also advised.

Storage Temperature

All liquid and gaseous chlorine should be stored at a temperature between 35°F and 120°F.

Compressed Gases

Compressed gases are used on construction sites for a variety of purposes. The most common uses include the generation of flame for welding and cutting metals. Compressed gases include relatively inert gases like air and carbon dioxide (used for blow cleaning or pressure testing), highly corrosive or reactive chemicals, (like chlorine and oxygen) and highly flammable gases (such as hydrogen and acetylene).

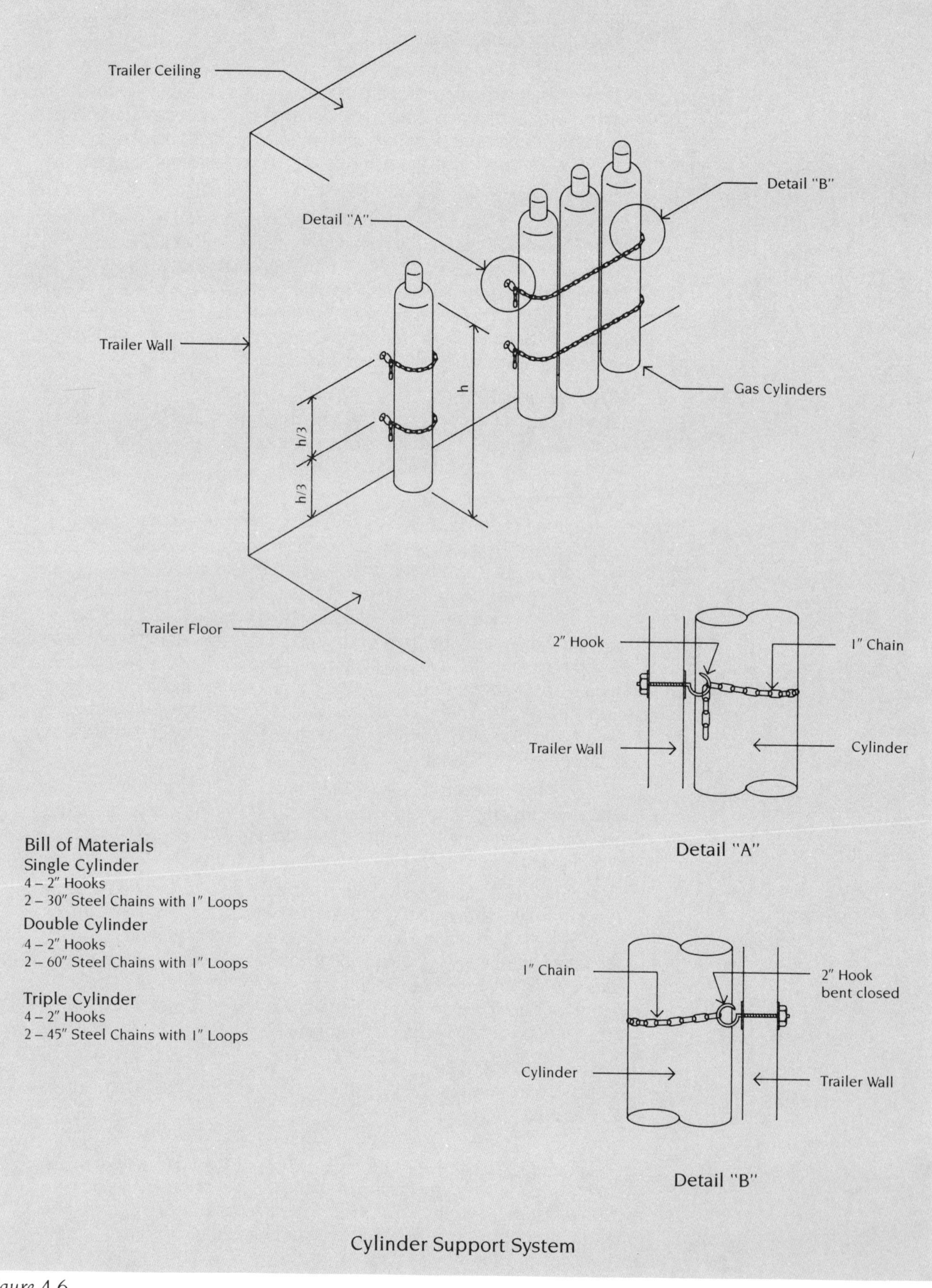
Trailer Ceiling
Detail "B"
Detail "A"
Trailer Wall
h
h/3
h/3
Gas Cylinders
Trailer Floor
2" Hook
1" Chain
Trailer Wall
Cylinder
Detail "A"
Bill of Materials
Single Cylinder
4 – 2" Hooks
2 – 30" Steel Chains with 1" Loops
Double Cylinder
4 – 2" Hooks
2 – 60" Steel Chains with 1" Loops
Triple Cylinder
4 – 2" Hooks
2 – 45" Steel Chains with 1" Loops
1" Chain
2" Hook
bent closed
Cylinder
Trailer Wall
Detail "B"
Cylinder Support System

Figure 4.6

Storage

Storage of gas cylinders must ensure protection from heat and physical damage, particularly to valves and gauges at the top of the cylinders. Gases that are normally used together must be segregated in storage. This means that, while most compressed gases can be safely stored together with other compressed gases, materials such as oxygen and acetylene must be stored separately. Those two gases should, in fact, be stored separately from all other gases due to their particularly aggressive nature. Incompatible substances should not be stored together in any case. A check of the Material Safety Data Sheets supplied by the gas distributor will help to identify incompatible substances.

Gas cylinders are generally stored in a vertical position because it is easier and safer for workers to pick up the cylinders in this position on a two-wheel truck to transport them than it would be if the cylinders were placed in a horizontal pile. Moreover, the vertical (rather than horizontal) alignment provides far greater protection for the valves and gauges located on the top of the cylinder. One drawback of vertical placement is that the cylinders are prone to falling over, an event which will almost always result in severe damage to the valves and gauges. The result of such damage may be leakage, with potentially lethal consequences. Even a small leak can cause a cylinder to shoot off like an uncontrolled missile, leading to serious damage to structures, or injury and potential death to people.

Storage, then, requires that cylinders be secured to a permanent wall or structure. This structure must be able to support the weight of all the cylinders secured to it. Attachment is generally accomplished with two steel chains pulled tightly around a group of several cylinders. The chains should be located approximately one-third and two-thirds of the way up the cylinder from the floor. A permanent fastener at one end of the chain prevents it from being lost, and a hook on the wall at the other end allows adjustable fastening to accommodate a varying number of cylinders of different sizes. Not more than five or six cylinders should be fastened with the same chain at the same time.

Figure 4.6 is a sketch of an appropriate fastening system for gas cylinders. Figure 4.7 shows a binned storage area for larger quantities of compatible gases.

Storage Temperature

All storage areas for compressed gases should be well-ventilated and maintained at a temperature not exceeding 120°F.

Coolants

Coolants are used to protect engines from overheating in heavy construction equipment, like bulldozers, trucks, backhoes, and cranes, and for trailer-mounted compressors and generators. Also in this category are refrigerants used in water chilling systems, air conditioning units, and other refrigeration units. Coolants are also generally included in antifreeze compounds. Typical examples include ethylene glycol (the most commonly used engine coolant and antifreeze), chlorodifluoromethane (a refrigerant commonly called *Freon* 22), and fluorotrichloromethane (a refrigerant commonly referred to as *Freon* 11).

Health Hazards

Since vaporization occurs at relatively high temperatures, coolants and refrigerants are generally toxic only when directly ingested. The systemic effects of coolants on humans can be severe, leading to kidney failure or

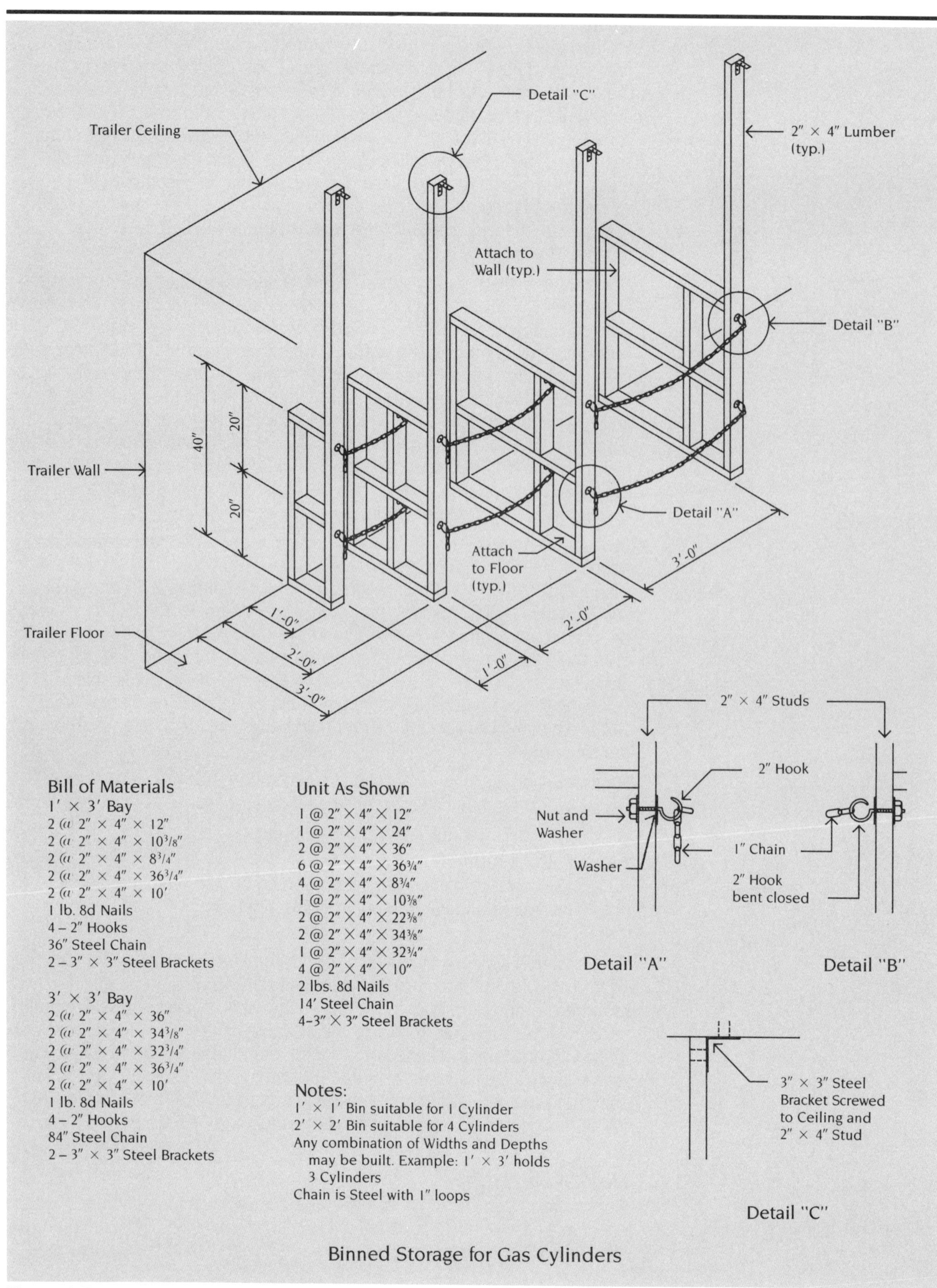

Trailer Ceiling
Detail "C"
2" × 4" Lumber (typ.)
Attach to Wall (typ.)
Detail "B"
Trailer Wall
40"
20"
20"
Detail "A"
Attach to Floor (typ.)
Trailer Floor
1'-0"
2'-0"
3'-0"
3'-0"
2'-0"
1'-0"
2" × 4" Studs
2" Hook
Nut and Washer
Washer
1" Chain
2" Hook bent closed
Detail "A"
Detail "B"
Bill of Materials
1' × 3' Bay
2 @ 2" × 4" × 12"
2 @ 2" × 4" × 10 3/8"
2 @ 2" × 4" × 8 3/4"
2 @ 2" × 4" × 36 3/4"
2 @ 2" × 4" × 10'
1 lb. 8d Nails
4 – 2" Hooks
36" Steel Chain
2 – 3" × 3" Steel Brackets
3' × 3' Bay
2 @ 2" × 4" × 36"
2 @ 2" × 4" × 34 3/8"
2 @ 2" × 4" × 32 3/4"
2 @ 2" × 4" × 36 3/4"
2 @ 2" × 4" × 10'
1 lb. 8d Nails
4 – 2" Hooks
84" Steel Chain
2 – 3" × 3" Steel Brackets
Unit As Shown
1 @ 2" × 4" × 12"
1 @ 2" × 4" × 24"
2 @ 2" × 4" × 36"
6 @ 2" × 4" × 36 3/4"
4 @ 2" × 4" × 8 3/4"
1 @ 2" × 4" × 10 3/8"
2 @ 2" × 4" × 22 3/8"
2 @ 2" × 4" × 34 3/8"
1 @ 2" × 4" × 32 3/4"
4 @ 2" × 4" × 10"
2 lbs. 8d Nails
14' Steel Chain
4-3" × 3" Steel Brackets
3" × 3" Steel Bracket Screwed to Ceiling and 2" × 4" Stud
Detail "C"
Notes:
1' × 1' Bin suitable for 1 Cylinder
2' × 2' Bin suitable for 4 Cylinders
Any combination of Widths and Depths may be built. Example: 1' × 3' holds 3 Cylinders
Chain is Steel with 1" loops
Binned Storage for Gas Cylinders

Figure 4.7

brain damage at high doses. Refrigerants can cause frostbite symptoms after a short exposure at the low operating temperatures typically encountered with refrigeration systems.

Storage

Storage of these substances is compatible with the general storage of construction materials. Care should be taken to protect containers from physical damage and leakage. Strong oxidizing agents and acids are generally incompatible with coolants.

Storage Temperature

Storage areas should be maintained at a temperature below 120°F.

Flammable Solvents

Flammable solvents are used for several purposes, such as paint thinning and removal, degreasing, and cement and glue removal. When not used directly, these types of solvents are also found as the principal active ingredient in commercial mixtures designed to accomplish the same purposes. In such cases, the solvents impart their hazardous qualities to those other compounds and mixtures. Examples include methyl ethyl ketone, ethyl alcohol, mineral spirits, toluene, xylene, and naphthalene, or naphtha.

Health Hazards

The greatest hazards associated with flammable solvents tend to result from their flammability. Use or storage near high heat sources, open flames or sparks (such as from welding and cutting operations) is extremely dangerous. Flammable solvents may also cause dermatitis and dry skin (with prolonged exposure), and are poisonous if ingested. Inhalation causes narcosis, a feeling of light-headedness, dizziness or euphoria, as well as headaches and nausea. Long-term exposure from ingestion or inhalation can lead to kidney and liver damage.

Storage

The objectives of a flammable solvent storage strategy are protection from heat and flames, protection from strong acids and strong oxidizers, and physical protection of the containers.

Small containers of flammable solvents, up to five gallons each, should be stored in fire-resistant steel cabinets such as those shown in Figure 4.8. The cabinet itself should be stored in a well-ventilated, dry area, out of direct sunlight. Larger containers and drums should be stored in a cool, dry location, also out of direct sunlight. Storage cabinets and larger containers must be stored away from engines, motors, and other sources of high heat, direct flames, and sparks. Many of these solvents are explosive over a wide range of concentrations in air, and any leakage or vapors from improperly cleaned spills can cause a rapid spread of fire.

When storing these materials in a trailer, the same general guidelines should be followed. Small containers should be stored inside metal cabinets within the trailer, with drums stored near the cabinet. A barrier to protect the drums from physical damage is recommended. See Figure 4.5 for a suggested protective barrier design.

Fuels—Gasoline/Diesel

These materials and their derivatives (such as kerosene) are most commonly used as fuels for motor vehicles, construction equipment, generators, compressors, chain saws, and space heaters. They include leaded and unleaded gasoline, white gas, No. 2 fuel oil, diesel fuel, and kerosene.

Health Hazards

Contact with the lighter fuels causes rapid drying of the skin, leading to

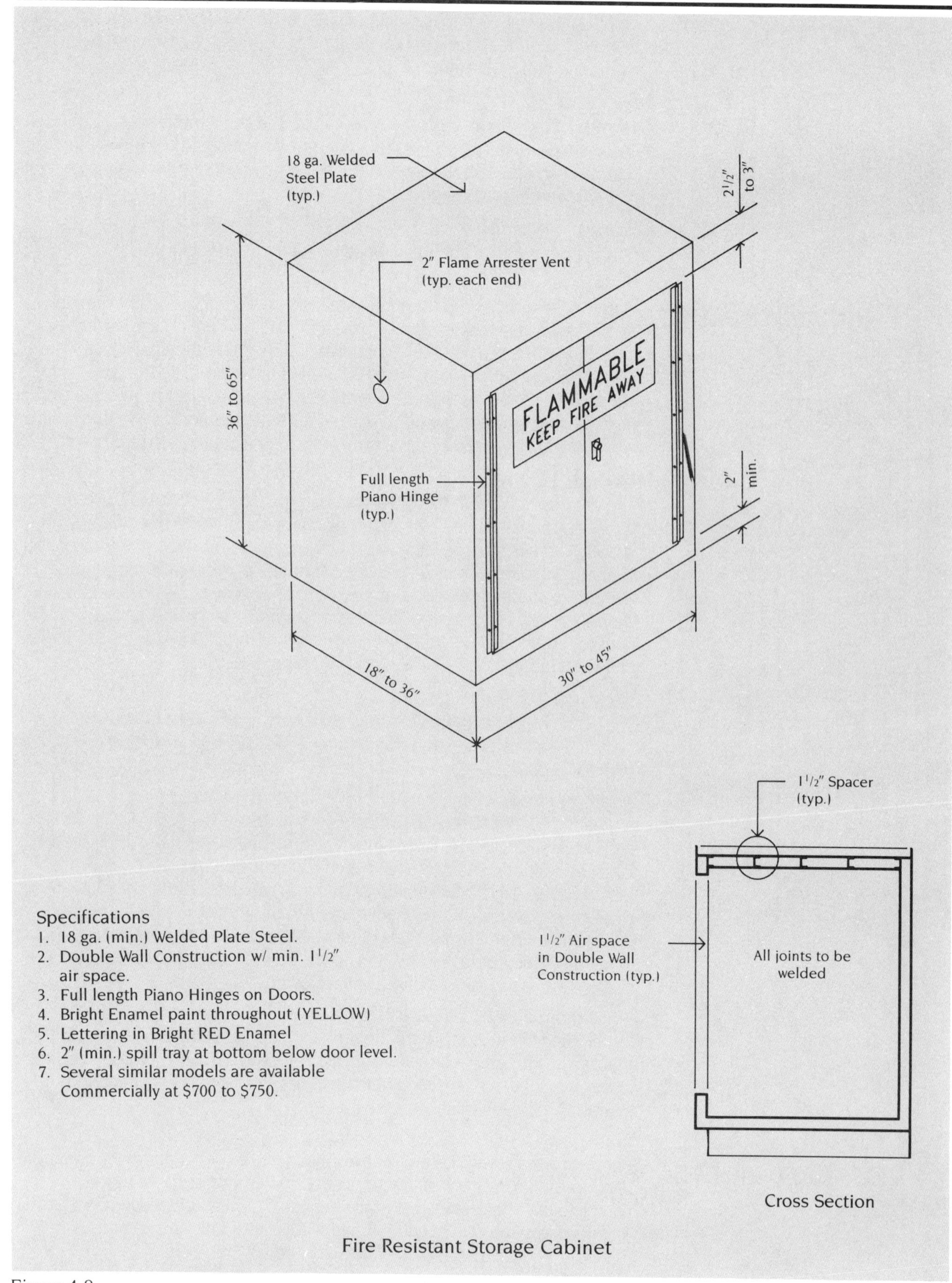

Fire Resistant Storage Cabinet

Figure 4.8

chapping, cracked skin, and dermatitis. The vapors are an irritant to the eyes, nose, and throat. Inhalation leads to dizziness, nausea, and headaches. Ingestion is poisonous, and causes damage to the central nervous system, kidneys, and liver.

Physical Hazards

The hazards associated with these compounds are generally related to their high flammability. Gasolines and kerosenes are explosive over a wide range of vapor concentrations in air, and they volatize at very low temperatures. Diesel fuels tend to be more difficult to ignite, except when vaporized or atomized, but burn at a higher temperature once lit.

Storage

Fuels generally arrive at the construction site in one of three ways: in a very large storage container (such as a trailer-mounted fuel pod of 200 to 500 gallons or more), in very small containers of one-to-five gallon capacity, and in the fuel tanks of motor vehicles and equipment.

Small fuel cans must be safety containers approved by the National Fire Protection Association (NFPA). Storage of those containers should be in fire-resistant steel cabinets, such as that shown in Figure 4.8. Minimizing the number of small cans on the construction site is strongly recommended. Cabinets may be stored in trailers or in other storage areas away from heat, flames, and sparks. Small fuel cans may be stored with paints and thinners. Acids are incompatible.

Trailer-mounted fuel pods are either moved to the equipment they are to service, or set in a fixed location to which the equipment returns for refueling. Whenever possible, a fixed refueling location is strongly recommended. A fixed location allows for much easier spill control than is possible with a mobile pod. Fixed refueling locations should provide diked spill containment, a concrete fueling pad, and fire protection equipment.

In any case, trailer-mounted fuel pods and tank trucks should be stored in a separate location away from buildings and other vehicles. They should not be stored indoors. An open-sided, roofed storage area is generally acceptable, subject to local building and fire code restrictions.

Paints and Thinners

Paints and thinners usually arrive on the job site in one gallon cans or plastic bottles, packed four or six to a box. Storage of unopened paints in their original shipping containers and cartons is strongly recommended. Cartons should be stacked on wooden pallets tall enough to prevent damage from minor flooding, and should be protected from precipitation. Stacking in trailers is acceptable.

Thinners are often shipped in five gallon metal cans or plastic bottles. Unopened containers can be stacked on wooden pallets in the same general area as paints.

Open containers of paints and thinners should be stored in a fire-resistant metal cabinet (painted yellow), similar to that shown in Figure 4.8. The cabinet may be kept in a general storage area or a trailer.

Buckets, brushes, pots, rollers, trays, and other painting supplies and equipment should be stored with the open paints and thinners in a fire-resistant metal cabinet.

Storage Temperature

Storage temperatures for paints and thinners should be maintained between 40°F and 120°F. Paints should be protected from freezing.

Phenolic Compounds

Phenolic compounds are typically used as sealants and coatings for waterproofing the outside of foundations, as cements and moisture barriers, as roofing materials, and as a wood preservative. Some examples are creosote, brick oil, carbolic acid, hydroxytoluene and phenol.

Health Hazards

Phenolic compounds tend to be corrosive to the skin and are tenacious when contact is made. The constituents are highly toxic. Absorption through the skin is the most common route of exposure. The fumes can be overpowering and cause gagging.

Physical Hazards

Phenolic compounds are incompatible with strong oxydizers, acids, and calcium hypochlorite.

Storage

Storage of these materials, which are usually shipped in five-gallon buckets, requires protection from physical damage, good ventilation, and isolation from heat or spark sources. Storage of buckets on wooden pallets, well away from acids, oxydizers, and hypochlorite, is recommended. Outdoor storage is preferred. Storage inside a trailer is discouraged, although sealed containers should pose no particular risk. Open containers should not be stored in trailers.

Storage Temperature

Storage temperatures should be maintained between 35°F and 120°F.

Oils and Lubricants

Oils and lubricants are used in motor pool operations, on heavy equipment and vehicles, and on compressors and generators. Oils are also applied to forms prior to placing concrete.

Health Hazards

Unused oils and lubricants are usually not dangerous, if properly handled. However, prolonged skin contact may cause irritation. Oils or lubricants mixed with solvent may cause eye and respiratory system irritation. They are generally toxic if ingested.

Storage

These materials are shipped to the site in all sizes of containers from 55-gallon drums down to one-pound grease gun cylinders. Containers can be steel, plastic, or cardboard, and the contents may be nearly pure petroleum products, such as motor oil and bearing grease, or a mixture of chemicals, such as transmission fluid or molybdenum-based grease.

Storage requirements are designed to protect the containers from physical damage. Small containers should be stored in their original shipping cartons. The cartons should be stacked on wooden pallets to protect them from minor flooding and stored under cover to protect the materials from precipitation. Drums should be stored near the pallets. Storage in a dry trailer is also acceptable.

Storage Temperature

In all cases, storage area temperatures for oils and lubricants should be maintained between 35°F and 120°F.

Figure 4.9 summarizes the storage requirements for all classes of chemicals and materials as outlined above.

Summary of Storage Requirements				
			Storage	
Chemical	Health Hazards	Physical Hazards	Physical	Temperature
Acids	Tissue burning, highly corrosive and toxic	Violent reaction with alkalais or water	Need special storage cabinet (See Figure 4.4)	35°F to 120°F
Alkalais	Tissue burning, corrosive, toxic	Violent reaction with acids, dust hazard	Keep dry, avoid damage	35°F to 120°F
Chlorinated Solvents	Dermatitis, Toxicity	Toxic fumes when burned	Keep away from acids, heat, fire	35°F to 120°F
Chlorine	Tissue burning, toxic	Highly corrosive to metals	Keep pallets dry, protect cylinders from damage	35°F to 120°F
Compressed Gases	Variable	Explosive pressures	Keep away from high heat, damage to tanks	Less than 120°F
Coolants	Frostbite, toxicity	Groundwater contamination	Keep away from acids, alkalais, protect from damage	Less than 120°F
Flammable Solvents	Dermatitis	Highly flammable	Keep away from heat, fire, sparks. Need special storage cabinet. (See Figure 4.8)	Less than 120°F
Fuels	Dermatitis, poisonous	Highly flammable, explosive	Safety cans and cabinets. Keep away from sparks, flames. (See Figure 4.8)	Less than 120°F
Paints & Thinners	Toxic fumes	Flammable, contaminate groundwater	Safety cabinets (See Figure 4.8)	40°F to 120°F
Phenolics	Corrosive to tissue, toxic	Incompatible with acids, calcium hypochlorite	Keep dry, away from heat, sparks, provide good ventilation	35°F to 120°F
Oils & Lubricants	Skin irritation	Sustain combustion	Protect from spills	35°F to 120°F

Figure 4.9

Chapter 5

Handling of Hazardous Materials

The improper handling of hazardous materials has very costly results. Personnel absenteeism, long-term personal injury claims, environmental pollution cleanup costs, and potential fines and legal fees associated with the improper handling of hazardous materials cost hundreds of millions of dollars every year. It does not have to be so. Not only does the proper handling of hazardous materials protect workers' welfare, but it is far easier, more economical, and more productive than improper procedures.

The proper handling of hazardous materials is founded on two primary concepts: **protection of personnel** and **protection of the environment**. If those two goals are met, the work site, the company, and the bank account will also be protected.

This chapter addresses the various types of protective equipment available for use on the job site when working with hazardous materials. Every employee should be familiarized with the equipment appropriate to the materials being used. Part of the *Right-to-Know* training required by law provides precisely that information. The employer is responsible for both providing the equipment and ensuring that employees use it.

This chapter links the appropriate protective equipment with the various types of hazardous materials. It is important to recognize, however, that it is not possible to cover in one book the appropriate protective equipment for every compound known to man, or even all those that are potentially available on a construction site. That is where the Material Safety Data Sheets, so often mentioned in this book, play a key role. There are MSDS's for every hazardous product, and each identifies the appropriate protective equipment to use when working with that product. Most MSDS forms tend to err on the side of safety, so their recommendations can be considered reliable.

This chapter also addresses spill prevention for various types of materials. The best cure for environmental damage is, of course, prevention. Being alert to the common sense measures outlined for spill prevention and containment minimizes the need for the information in Chapter 6, "Spill Cleanup Procedures."

Types of Personal Protective Equipment

Wherever any employee is working with a hazardous material or a hazardous waste, use of personal protective equipment (PPE) should be mandated by standard company policy. The basic concept of using PPE is often misconstrued. PPE is *not* for the exclusive use of hazardous waste cleanup contractors. More importantly, it does not require the use of "moon-suits" (fully-encapsulating suits) during normal construction activities.

For most construction activity, except work with asbestos and PCB's (as outlined in Chapter 7), PPE consists of gloves, overalls, boots, aprons, air purifying masks, ear muffs, and eye protection, or some appropriate combination thereof. The key is to recognize what each type of equipment does and does not do. An appropriate ensemble can then be selected.

The following section reviews some of the important characteristics of the various types of personal protective equipment appropriate to a job site. It is *not*, nor is it intended to be, an all inclusive dissertation on the subject of PPE use. Before beginning work with any new substance, an employee should carefully review the appropriate MSDS form and select the PPE recommended on the MSDS. A chart summarizing this analysis is presented in Figure 5.1.

Head Protection

Hardhats made of plastic, rubber, or other OSHA-approved material, are standard equipment on all job sites. They protect the head from impact injury. Hardhats are often provided with liners to insulate against cold. Liners are not, however, adequate protection against chemical splashing. Where splash protection is needed, during spray washing of wall surfaces for example, a chemically-resistant hood should be used over or under the hardhat.

Eye and Face Protection

The lowest level of face and eye protection is a sweat band. While not usually characterized as PPE, a sweat band helps to prevent sweat-induced irritation of the eyes and consequent vision impairment. A sweat band is particularly useful in a high heat environment, where irritant dusts or mists are present.

Safety glasses are a commonly used eye protection device. They protect the eyes from large particles of dust and from projectiles. Note, however, that safety glasses do not fit tightly against the face. That means that chips, projectiles, and dust can go around or behind the glasses and still damage the eye. Safety glasses are a sound protective measure where the possibility of damage exists, but the probability of direct exposure is low. Safety glasses should be fitted with special protective lenses and required on the job site when lasers are being used for site survey and alignment work.

Safety goggles can provide a much higher level of personal protection than safety glasses because goggles fit tightly against the face. Depending on their construction, goggles can provide protection against vapors and mists, and chemical splashes. If high impact-resistant lenses are used, goggles can also protect against large particles and projectiles. High impact-resistant goggles are recommended for work involving grinding, cutting, drilling, power sawing, chipping, and other such activities where large particles may fly off from the work.

When working with liquids, such as chemical sprays with high-pressure hoses and steam jennies, where splashing is likely but projectiles are not, use of a full face shield or splash hood is recommended. Splash hoods protect the entire head and neck from splashes, while a full face mask protects only the face area. Neither device is suitable for protection against projectiles, vapors, dust, or mist. Use of safety glasses or goggles, in conjunction with either a full face mask or splash hood, is recommended where the operation may also cause projectiles.

Personal Protective Equipment (PPE)		
PPE	Protection Provided	Limitations
Hard Hat	Impact injury to head	No protection against chemical splashing, gases, fumes, vapors, or mists
Hood	Chemical splashing	No protection against impact, gases, mists, fumes, or vapors
Sweat Band	Minimal protection of eyes from perspiration-borne irritants	No real protection of any kind provided
Safety glasses	Large dust particles, chips, and projectiles. Laser protection with proper lenses	Do not fit tightly, no protection from dust clouds or side impact projectiles
Goggles	Dust, projectiles, chips	Not good for liquid splashes, gases, or mists
Full Face Mask	Higher projectile protection, good against splashes and sprays	Need hood for full protection against splashes and sprays. Need respirator to protect against mists, gases, vapors or fumes
Ear Plugs	Noise	Inconvenient to use
Ear Muffs	Noise	Bulky to wear (but improving)
Apron	Splashes and spills, good for working with acids, corrosives, etc.	Only cover clothes in certain areas; no protection of outside edges
Coveralls	Dust, spills	Not good for liquids, gases, vapors, or fumes
Rubberized Gear	Liquid sprays, mists, and spills	Bulky, hot to wear, cumbersome to work in
Gloves	Hand protection from chemical contact, specialty gloves for high heat, etc.	Only protect to top of glove. Material must be compatible with chemical being handled
Sleeves	Increase protection from gloves	May give false sense of security
Special Shoes	Steel-toed protect against crushing injury to toes. Some provide puncture protection. Special shoes provide chemical protection	Not all shoes are the same. Special shoes needed for special situations
Boots	Protect feet from chemical spills and splashes	Must use care to avoid spills to inside of boot
Dust Mask	Good general protection from dusts	Not effective against gases, mists, vapors, or fumes
SCBA	Excellent against all types of air-borne contaminants	Air supply is self-contained, limiting work time
SAR	Excellent against all types of air-borne contaminants	Mobility limited by air supply hose
APR	Excellent against air-borne contaminants for which designed	Special filters needed depending on chemical in atmosphere

Figure 5.1

Ear Protection

Ear protection is, fortunately, becoming increasingly common on construction sites. Ear plugs are often provided to casual site visitors or may be used infrequently by workers. The inconvenience of removing the plugs to hear direction or instructions and then having to replace them to resume work, is a disincentive to employees who should wear them. In addition, each time ear plugs are removed and reinserted on a work site, the possibility of direct ear damage increases.

Ear muffs have begun to replace ear plugs for most applications. The bulky mass of ear muffs would initially appear to be a disadvantage. However, they are now constructed of lighter-weight materials that fully protect against hearing loss, while still being acceptably comfortable to wear. Moreover, they can be easily shifted to one side so that a command or instruction can be heard. All in all, ear muffs provide good protection at low risk.

All workers conducting any operation that produces high volume or high intensity sound should be required to wear ear protection. Activities that typically fit this category are grinding, power sawing, ram-setting, concrete finishing, machine operation, concrete sawing, power equipment operation, heavy equipment operation and maintenance, pile driving, and sanding. In general, if a noise prohibits normal conversation or causes discomfort, ear protection should be required.

Body Protection

Body protection in the context of construction site operations means protection from projectiles or splashing of chemicals. Included in this category are blast and fragmentation suits only for those employees normally engaged in blasting operations. It does not include fully-encapsulating suits ("moon-suits") or other extreme protective measures except for those workers who are directly involved in and properly trained to conduct hazardous material cleanup operations.

The most common body protective gear is a simple apron. Usually made of neoprene, or neoprene-coated canvas, aprons cover the entire body from the neck to below the knees. Aprons are easy to don and provide good protection from splashes and spills for those working with hazardous chemicals. Aprons are particularly useful when working with battery acids, strong alkalis, or corrosives, or when mixing cleaning solutions.

Coveralls are commonly used to provide protection against greases and oils in vehicle maintenance areas. They also offer protection against the dust of powdered forms of chemicals. Rubberized rain gear or similar types of jackets, hoods, and pants are useful when working with spray operations. It is important to verify the compatibility of the protective gear material with the chemical being used. That information comes from the material MSDS. Note that coveralls and rain gear-type garments do not protect against gases, vapors, or fumes.

Hand and Arm Protection

Gloves and sleeves are used to protect hands and arms. Neoprene or rubber gloves, depending upon the recommendations found on the pertinent MSDS, should be worn when dealing with acids, alkalis, solvents, and degreasers or other harsh chemicals. It is not uncommon to find employees carelessly dunking hands and forearms into vats of Freon, trichlor, and other solvents when cleaning parts and tools. Such practices are, however, extremely risky and should be prohibited by an established and enforced company policy.

Asbestos or heat-resistant gloves are worn by welders, metal grinders, and others working near high heat. Sleeves or long-sleeved gloves should be used by welders, grinders, and those working with acids, alkalis, or abrasive chemicals to avoid burns from hot sparks or chemical splashing.

Foot Protection

There are two types of foot protection to consider on a construction site. One is for the prevention of impact or crushing damage, and the other is to prevent chemical damage. Most construction workers tend to wear heavy duty boots as a matter of habit. Boots that combine steel shanks, toes, or insoles protect against crushing, compression, and puncture injuries. However, only those boots constructed of special chemical-resistant materials will provide protection from chemical spills.

In situations where feet are continously exposed to hazardous chemicals (particularly liquid chemicals), the use of rubber or neoprene boots is recommended. Disposable booties to fit over regular construction shoes are available for infrequent or short-term use. Tall, fitted boots are recommended for long-term exposure.

Respiratory Protection

Respiratory devices seem to create more unjustified confusion and mystery than all the other PPE combined. This need not be so. In fact, most activities on a construction site require no respiratory protection at all, and those that do, usually require only minimal protection.

Most of the confusion surrounding respiratory protection is due to the wide variety of commercial respirators that are available. Except when working with asbestos (see Chapter 7), respirators, per se, are not usually required on construction projects. Dust masks are often used, however, and with good cause.

Respirators come in three basic types:

1. Self-contained breathing apparatus (SCBA), which supplies air from a source carried by the user.
2. Supplied-air respirators (SAR's), which supply air from a remote source via a hose connected to the respirator.
3. Air-purifying respirators (APR's), which purify ambient air through a filter element before it is inhaled.

Special training is required to use SCBA's and SAR's. Moreover, APR's are limited to specific chemicals, as determined by the nature of the filter media: their duration of use is limited, and they must be used in areas that have an adequate oxygen supply. Because of the cited limitations, use of respirators on the job site should be considered an emergency practice, not a standard practice.

Given the limited use of respirators on the job site, only dust masks are used as a regular practice. Dust masks are available in a variety of shapes and sizes, but all provide the same basic function of filtering dust particles out of the air inhaled by the wearer. Dust masks provide only minimal vapor and fume protection and virtually no protection against gases. It is important, therefore, to ensure that all areas subject to gas, vapor, or fume accumulation be well-ventilated at all times.

Dust masks should be used when working with dry solids, dry chemicals, sanding equipment, sawing equipment, and sand-blasting equipment, and when conducting other operations that are prone to dust generation. Most filters are made of paper for easy disposal and should be changed frequently in areas of heavy exposure. Whenever breathing becomes difficult inside the mask, or the exterior surface becomes visibly clogged with dust, the user should move to a dust-free area and change the filter.

Emergency Protection

It is occasionally necessary to work in confined spaces on a job site and accidents do occur. Unfortunately, people tend to respond to emergencies without carefully thinking through their actions, particularly if there is no pre-planned emergency response protocol, equipment, or personal protective gear. More would-be rescuers die each year trying to save accident victims than the number of original victims. It is important, therefore, to pre-plan emergency response procedures and to provide adequate safety equipment – and the personnel training to go with it – to allow for a safe rescue.

When a victim succumbs to toxic fumes or a lack of oxygen, it makes no sense to send more people into the hazardous atmosphere unprotected. Dust masks are not suitable for toxic or oxygen-deficient atmospheres. Only SCBA's and SAR's (self-contained breathing apparatus, and supplied-air respirators) should be used in such cases. Unless the nature of the toxin is clearly identified, it is positively known that there is sufficient oxygen in the atmosphere, and an appropriate APR (air-purifying respirator) is readily available, SCBA's are recommended.

Use of SCBA equipment by untrained personnel is dangerous. Only properly trained safety personnel should attempt a rescue from a contaminated atmosphere. The appropriate type of SCBA for a given job site should be determined in conjunction with the development of an emergency response plan prepared by the site safety officer. If no safety officer is on the site, SCBA should not be used to enter a toxic or contaminated atmosphere.

Recommended Uses of Protective Equipment

The sections that follow describe the specific types of personal protective equipment recommended for use when working with particular types of material. In all cases, the MSDS supplied by the chemical manufacturer should be consulted before working with hazardous materials. Figure 5.2 is a chart summarizing this information.

Acids

Acids burn skin and clothing on contact. Therefore, the appropriate protective equipment is designed to prevent contact in the event of minor spillage or splashing.

Personal Protective Equipment

Eye protection is best afforded by wearing safety goggles or a full face mask. For the occasional use of small quantities of acids, such as topping off an automotive battery, goggles are appropriate. When using large quantities, or when brushing or spraying acids, such as when cleaning brickwork or brass, full face masks are appropriate. Neoprene gloves and aprons should also be used when handling small quantities of acids. Boots should be added where the washing of horizontal surfaces is involved. For spraying or brushing acid on walls, or power washing after acid application, aprons or neoprene coats should be worn in addition to the other gear noted.

Safe Work Practices

When working with small containers of acids, a box of common baking soda (bicarbonate of soda) should be kept within reach to pour generously on spills. The baking soda quickly neutralizes the acid and absorbs the liquid, thereby minimizing the resulting damage. After the soda/acid mix has stopped fizzing, it can be cleaned up safely with water and towels.

Where a large container of acid is maintained at a fixed location and transfers are made to smaller containers for transport to the point of use, the spout of the storage container should have a drip pan under it to catch drippings and

PPE Recommendations

Hazard	Hard Hat	Hood	Goggles	Full Face Mask	Ear Muffs	Apron	Cover-alls	Rain Gear	Gloves	Sleeves	Special Shoes	Boots	Dust Mask	Respirator
Acids		• (Spray)	•	• (Spray)		•		• (Spray)	•	•		• (Liquid)		
Alkalis		• (Spray)	•	• (Spray)		•	• (Dry)	• (Spray)	•				• (Dry)	
Chlorinated Solvents			•	• (Fumes)		•			•	•				• (Fumes)
Chlorine (liquid)			•			•			•	•				• (Mixing)
Chlorine (dry)			•			•			•	•			•	
Chlorine (gaseous)			•						•					• (Confined Spaces)
Compressed Gases			•		•				• (Not Rubber)		• (Steel Toed)			• (Spill Control)
Coolants			•						•	•				
Flammable Solvents			•			•			•	•				• (Confined Spaces)
Fuels						•	•		•	•	• (Chemical Resist.)			
Metals	•		•	• (Welding, Heat Cutting)	• (Sawing, Grinding)	• (Welding, Heat Cutting)			•		• (Steel Toed)			• (Confined Spaces)
Paints and Thinners	• (Cap)		• (Overhead Work)				•	• (Spray)	•					• (Confined Spaces)
Phenolics	• (Cap)		•				•	• (Spray)	•					• (Confined Spaces)
Oils and Lubricants			• (Sprays)				•		•		• (Chemical Resist.)			
Wallboard	•		•		• (Sanding)		•		•				• (Sanding)	
Wood Products	•		•		• (Sawing, Sanding)		•						•	

Figure 5.2

small spills. A one-to-two inch layer of baking soda on the bottom of the drip pan will effectively neutralize any drippings on contact. Figure 5.3 shows an appropriate arrangement for jugs and bottles. Figure 5.4 shows acceptable arrangements for a spigotted drum. A single drip pan for multi-tiered drum racks is suitable. The pan should be large enough, however, to totally contain the entire volume of liquid in the largest drum on the rack, plus 10%. Figure 5.5 shows the construction of a suitable drum rack. Figures 5.6 and 5.7 show the details of multi-tier drum racks and a multi-drum drip pan.

Alkalis

In contact with body tissue, alkalis burn in a corrosive manner. Eyes and mucous membranes are immediately affected, while skin tends to burn so slowly that no immediate effect is noticed. Inhalation burns the nose and throat, and ingestion destroys digestive tract tissue. Handling practices are directed toward prevention of contact with liquids or powders, the prevention of dust inhalation, and protection from liquid splashing.

Alkalis that are shipped in dry form and stored on pallets are often mixed with water before use. It is important to add the alkali to the water and then to stir gently. Do not add water to alkalis.

Personal Protective Equipment

When handling dry alkalis, aprons, gloves, goggles, and a dust mask are essential. Safety glasses are inadequate to protect eyes from alkali dust. When handling liquid alkalis, the dust mask is unnecessary. When spray cleaning with an alkaline solution, use of a full face mask, hood, and rain gear is recommended to avoid mist accumulation on the hair, face or scalp.

Safe Work Practices

Dry absorbent should be maintained near all liquid alkali storage areas for spill control. When mixing small quantities (less than five gallons), no containment device is required. If larger quantities are prepared for transfer to smaller containers as the need arises, or if bulk liquid alkalis are being used, the larger container should have a catch pan under it large enough to contain a spill of the entire container contents. Figures 5.4 and 5.5 show appropriate catch pan arrangements for spigotted drums of alkali. Note that acids and alkalis should not be stored together on multi-tier racks.

Chlorinated Solvents

The risks associated with chlorinated solvents on a construction site come from direct contact in cleaning vats, fumes from use on hot engines and parts, and fumes resulting from spray cleaning operations. Protective equipment should be selected to prevent contact or inhalation.

Neoprene gloves should be used whenever parts are immersed in solvents for cleaning, or when brushing solvents onto parts. Where hot parts are being cleaned, good ventilation is required to reduce fume accumulation, and goggles are necessary to protect eyes. If continuous or frequent exposure is likely, an appropriate air-purifying respirator should be used. Note that dust masks do not provide adequate protection against fumes, and safety glasses do not protect eyes from fumes or gases.

Chlorinated solvents should not be used in confined areas. If it is necessary to do so, a self-contained breathing apparatus (SCBA) or a supplied-air respirator (SAR) is required. Only properly trained personnel should don either of these types of respirators. If the material is being sprayed, a hood, full-face mask, gloves, rain gear, and SCBA or SAR should be used.

Chlorinated solvents tend to evaporate quickly if spilled. Nevertheless, evaporation should not be relied on for spill cleanup. A supply of dry

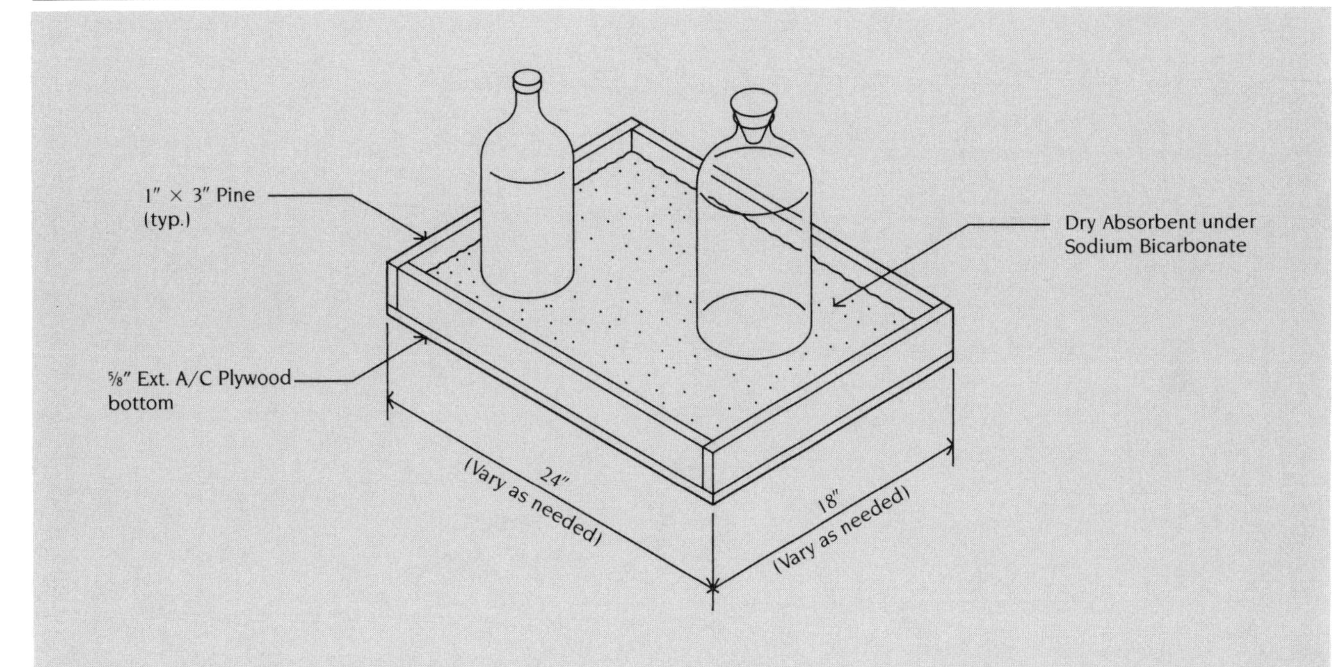

Note:
Cover bottom with 1" Dry Absorbent and
1/4" Baking Soda (Sodium Bicarbonate).

Bill of Materials
2 @ 18" × 1" × 3" Pine
2 @ 22 1/2" × 1" × 3" Pine
1 @ 24" × 18" × 5/8" A/C Ext. Plywood
1/2 lb. 6d Galvanized Nails
10 lbs. Dry Absorbent
1 lb. Sodium Bicarbonate

Bottle and Jug Drip Pan

Figure 5.3

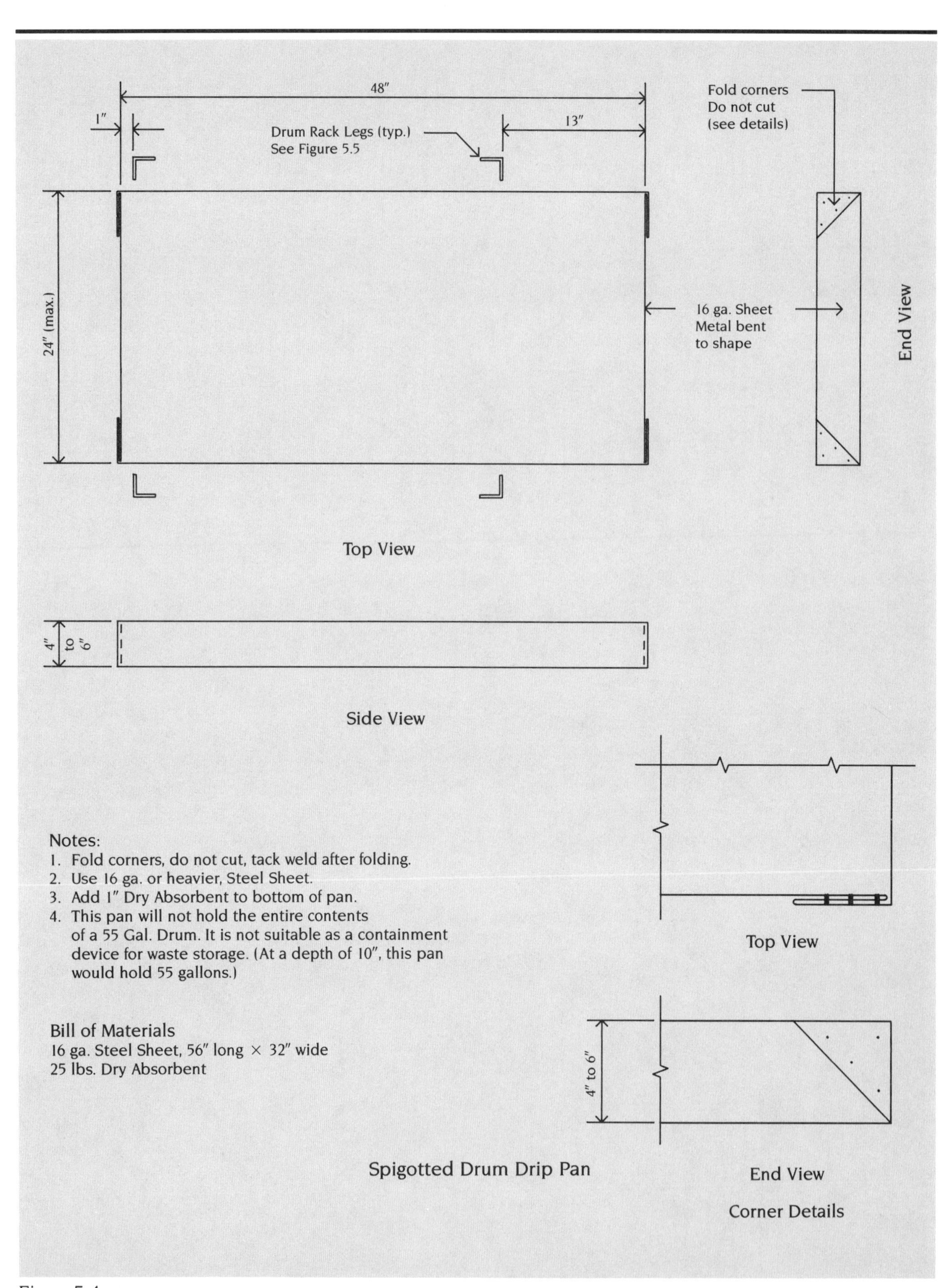
48″
1″
13″
Drum Rack Legs (typ.)
See Figure 5.5
Fold corners
Do not cut
(see details)
24″ (max.)
16 ga. Sheet
Metal bent
to shape
End View
Top View
4″ to 6″
Side View
Notes:
1. Fold corners, do not cut, tack weld after folding.
2. Use 16 ga. or heavier, Steel Sheet.
3. Add 1″ Dry Absorbent to bottom of pan.
4. This pan will not hold the entire contents of a 55 Gal. Drum. It is not suitable as a containment device for waste storage. (At a depth of 10″, this pan would hold 55 gallons.)
Bill of Materials
16 ga. Steel Sheet, 56″ long × 32″ wide
25 lbs. Dry Absorbent
Top View
4″ to 6″
Spigotted Drum Drip Pan
End View
Corner Details

Figure 5.4

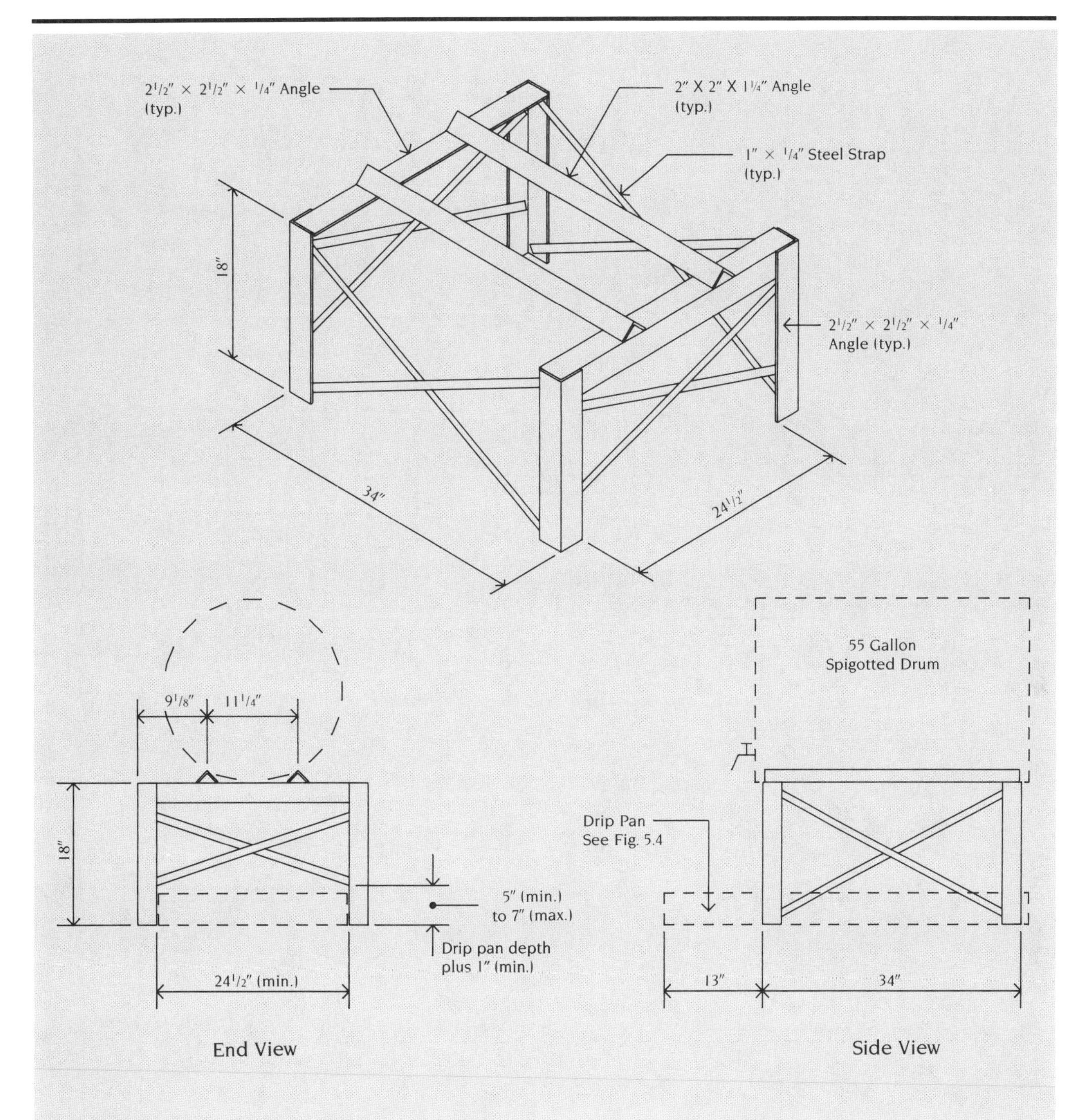

Bill of Materials

4 @ 2 1/2" × 2 1/2" × 1/4" × 18" Long Angle
2 @ 2 1/2" × 2 1/2" × 1/4" × 29 1/2" Long Angle
2 @ 2" × 2" × 1/4" × 34" Long Angle
4 @ 1" × 1/4" × 37 1/2" Long Strap
4 @ 1" × 1/4" × 29 1/2" Long Strap

Notes:

1. All members to be A36 Steel.
2. Weld all intersections and joints.

Single Drum Rack

Figure 5.5

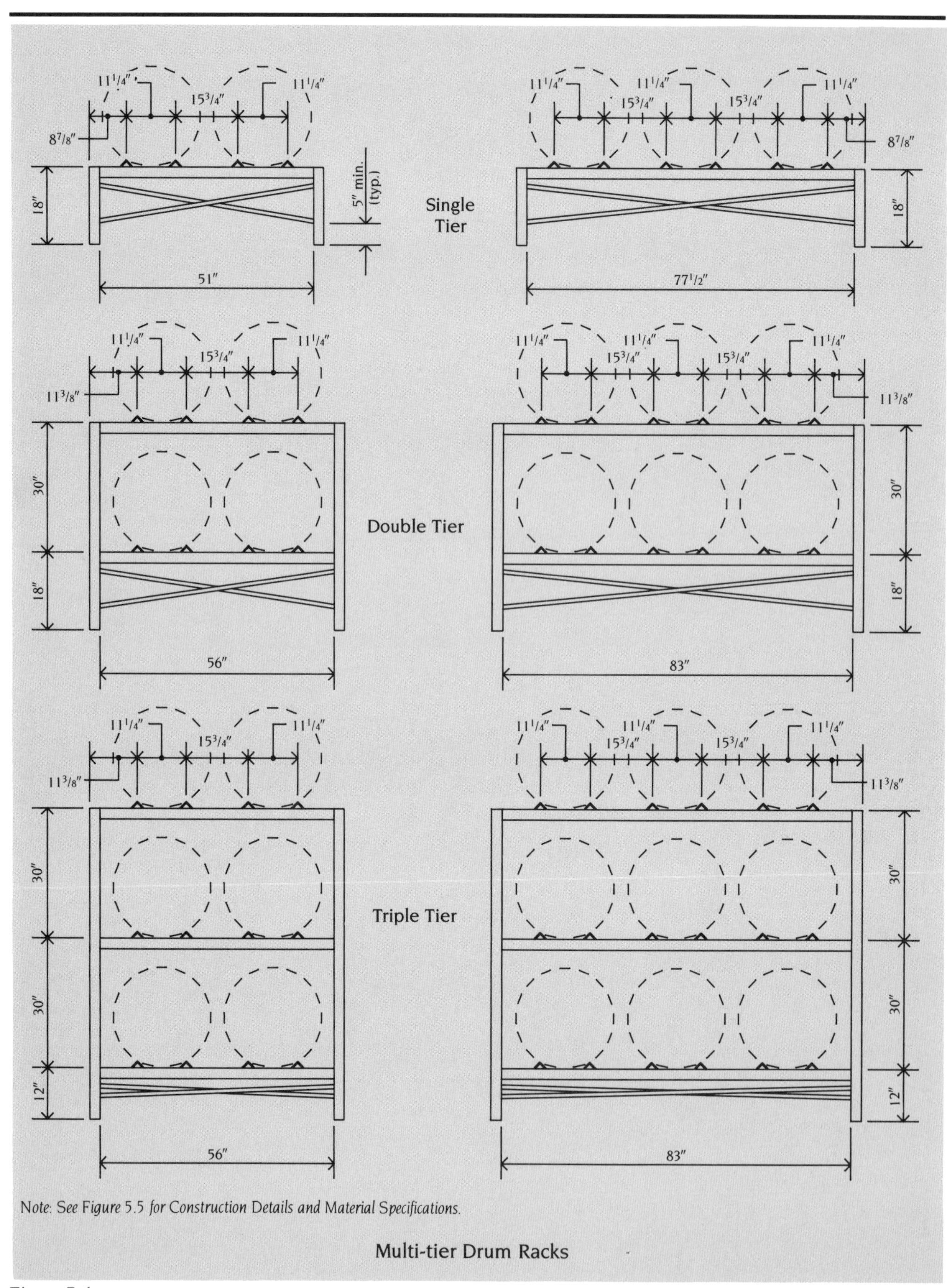

Note: See Figure 5.5 for Construction Details and Material Specifications.

Multi-tier Drum Racks

Figure 5.6

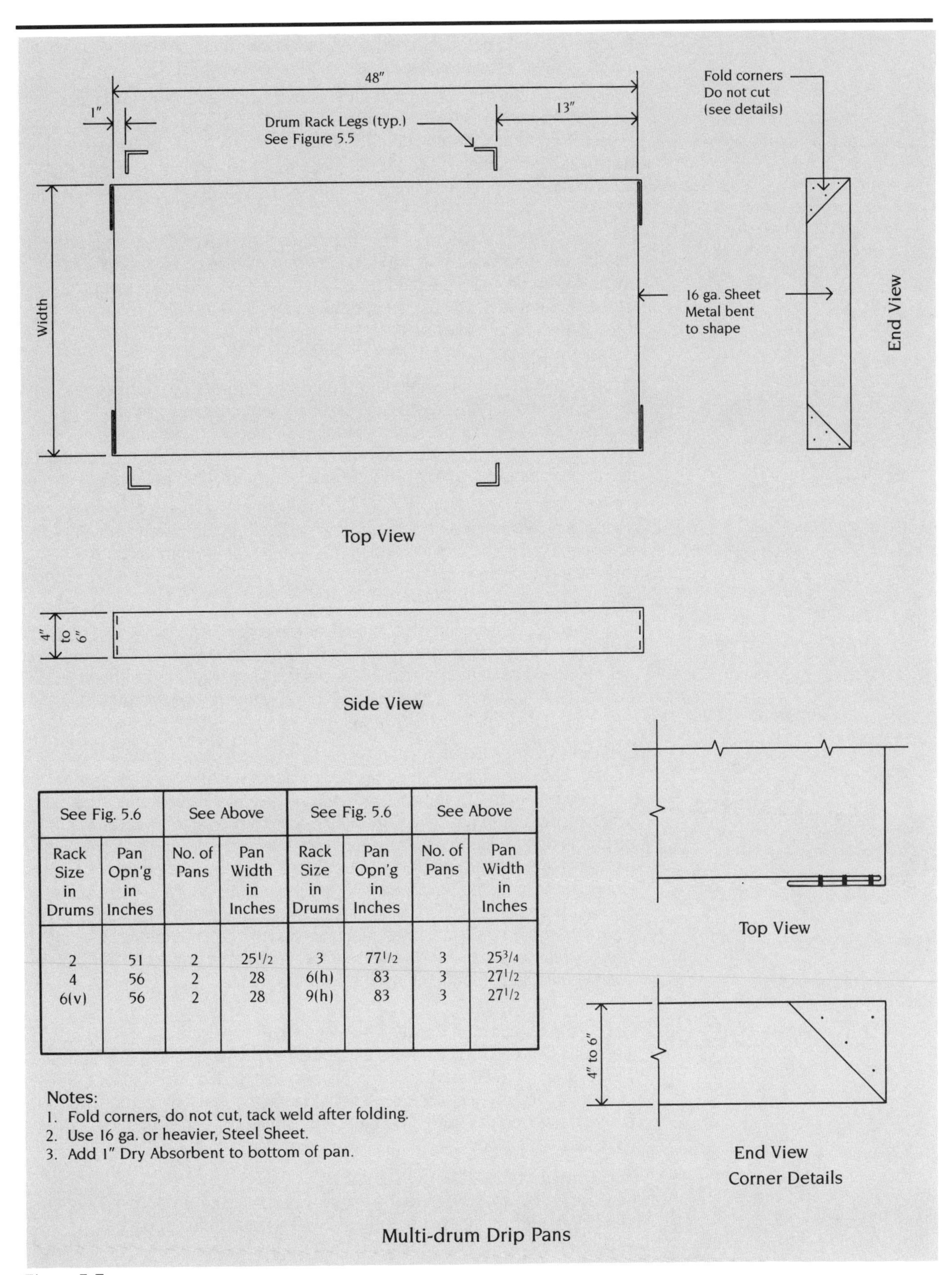

See Fig. 5.6		See Above		See Fig. 5.6		See Above	
Rack Size in Drums	Pan Opn'g in Inches	No. of Pans	Pan Width in Inches	Rack Size in Drums	Pan Opn'g in Inches	No. of Pans	Pan Width in Inches
2	51	2	25 1/2	3	77 1/2	3	25 3/4
4	56	2	28	6(h)	83	3	27 1/2
6(v)	56	2	28	9(h)	83	3	27 1/2

Notes:

1. Fold corners, do not cut, tack weld after folding.
2. Use 16 ga. or heavier, Steel Sheet.
3. Add 1" Dry Absorbent to bottom of pan.

Figure 5.7

absorbent should be maintained nearby whenever these solvents are used. In the event of spillage, the absorbent can be quickly applied.

Drums of chlorinated solvents should be fitted with tightly closing spigots. Figures 5.4 and 5.5 show appropriate drip pans for spigotted drums of solvent. Note that solvents should not be stacked with acids, alkalis, or other incompatible materials on multi-tier racks. Drip pans should be adequately sized to fully contain the contents of an entire drum.

Chlorine

The highly corrosive nature of liquid chlorine, the speed with which gaseous chlorine overpowers and kills, and the corrosive nature of dry hypochlorites under slightly damp-to-liquified conditions, dictate the handling protocols for these materials. The basic requirements for handling chlorine in each of these forms are listed below.

Liquid Chlorine

Chlorine cannot be maintained as a liquid when exposed to normal atmospheric conditions. Any that escapes its container very quickly evaporates to a deadly gas at temperatures well below those of normal atmosphere. Chlorine gas is easily dissolved in water, however, to create a strong oxidizing solution that is quite stable at normal temperatures.

For handling pre-mixed chlorine solutions, neoprene gloves, aprons, and eye goggles are recommended. If mixing gaseous chlorine into water, an air purifying respirator is recommended due to the high probability of free chlorine gas escaping.

Dry Chlorine

When mixing dry hypochlorites into water (never add water to a powder), gloves, aprons, eye goggles, and dust masks are recommended. If inhaled, hypochlorite dust damages respiratory tract passageways, hence the need for a dust mask. However, the likelihood of free chlorine gas being emitted in significant quantities is extremely small.

Gaseous Chlorine

Gaseous chlorine, shipped in compressed gas cylinders of various sizes, is extremely potent in a confined space. Mixing gas into water should only be done outdoors or in a very well-ventilated area. Gloves and eye goggles are recommended for all persons who are either working with the gas or working in a trench near the point of use. Those actually making the piping connections should be equipped with appropriate air purifying respirators. Respirators are essential in confined spaces. A self-contained breathing apparatus should be maintained in close proximity to chlorine gas use and storage areas, but not in the same room. An employee trained in using SCBA should be available during all such operations.

Compressed Gases

There are two important dangers associated with the handling of compressed gases. The first comes from physical damage to the gas cylinder, which would allow leakage of the hazardous contents or cause the cylinder to shoot off like an uncontrolled missile to cause serious damage or personal injury. The second danger comes from a possible leak of cylinder contents due to a faulty tank connection.

Personal Protective Equipment

It is strongly recommended that a two-wheeled truck be used for moving cylinders. The truck should be designed with a curved back specifically to hold cylinders, and a strap to attach them to the truck frame. Properly-stored cylinders can be easily picked up and safely moved by one person using such a device. Steel-tipped shoes or boots are suggested for workers who routinely move cylinders due to the weight of the truck and cylinders, which is sufficient to crush toes.

Because the contents of gas cylinders are often explosive in air, static electrical discharge from a worker to a cylinder should be prevented. Generally, proper work boots will insulate the worker, but use of leather or cloth gloves when working with or around gas cylinders is also recommended. Rubber gloves should not be worn, however, because the surface of a rubber glove is a good conductor of static electricity.

Safe Work Practices

When connecting a hose or tool to a gas cylinder, the threaded connectors must be carefully aligned to avoid cross-threading, and the resulting leaking connection. After the connection is made, a thin coating of oil should be placed all around the joint. Then when the valve is *slowly* opened, bubbles can be seen at the joint if it is not tight. If bubbles appear, the valve should be closed at once and the joint remade. If these precautions are taken, protective gear is not required, except that suggested by use of the tank contents in confined spaces.

Coolants

Coolants are usually toxic only when directly ingested. Vaporization is not a major concern except in confined spaces at elevated temperatures, or when handling compressed, gaseous refrigerants. Note, however, that refrigerants will cause frost-type injuries if liquid forms contact skin, and that refrigerant vapors can cause narcosis, or unconsciousness at low concentrations and death at higher concentrations.

Personal Protective Equipment

Personal protective equipment should include rubber gloves to guard against spills on the hands. Eye goggles should be worn while pouring liquids, to protect against splashing. In confined areas and elevated temperature zones, air-purifying respirators are strongly recommended. A review of the Material Safety Data Sheet for the coolant being used will help to identify the specific type of APR (air-purifying respirator) required.

Flammable Solvents

Flammable solvents evaporate rapidly to dry skin which it has contacted, and the vapors are toxic if inhaled. In addition, spills leave flammable residues.

Personal Protective Equipment

Rubber-lined gloves and rubberized aprons should be used when working with flammable solvents. Solid rubber gloves should not be used, as they transmit static electricity which could ignite the solvent. Eye goggles are also recommended. Use of solvents should be restricted to well-ventilated areas to avoid concentration of fumes and vapors.

If flammable solvents must be used in a confined area, an appropriate air purifying respirator or self-contained breathing apparatus is essential.

Safe Work Practices

While a "no smoking" policy may seem like an obvious requirement in an area where solvents are used, it is essential that all such areas be posted as "no smoking" areas, and that the ban be very strictly enforced.

Fuels—Gasoline/Diesel

Personal Protective Equipment

Gasoline and diesel fuel should be handled with rubber-lined gloves. Workers assigned full-time to refueling operations might also find rubberized aprons to be of value. Solid rubber gloves can carry static electricity, however, which could ignite fuel vapors, and they should not be used.

Safe Work Practices

Refueling operations should always be conducted in very well-ventilated areas, away from heat, sparks, and open flames. Vehicles and equipment should be shut off before the fuel tank is opened and should not be restarted until fueling is complete, the fuel tank is closed, and any spills have been cleaned up.

While obvious to most, it is important to note that all areas where refueling takes place must be posted as "no smoking" areas, and that the ban be strictly enforced.

Metals

Metals used on construction sites are not inherently dangerous materials. However, when metals are cut, ground, or welded, dust and fumes are released which can be hazardous to health.

Personal Protective Equipment

Sawing, grinding, and sanding operations require eye goggles and dust masks for personal safety. Since each of these activities causes the metal to get hot, leather or asbestos gloves should be used to handle the metals.

Torch-cutting and welding operations use extreme heat to melt metal. Hot sparks and droplets of molten metal fly off from the work area. Workers using torches or welding machines require special, dark glass face masks to protect eyes from the very intense light generated by the activity and to protect the face from burning sparks. A heavy rubber or leather apron is also required to protect the body from sparks. Long-sleeved, heavy leather or asbestos gloves are needed to move the work and to handle the tools. A hard hat is recommended to protect the scalp and hair from flying sparks. Steel tipped boots are recommended due to the high potential for large metal pieces falling over or dropping onto exposed feet.

When cutting or welding in enclosed spaces, the fumes from the molten metal and welding rods can accumulate quickly and suffocate a worker. An air purifying respirator (APR) is required when working in an enclosed area. The Material Safety Data Sheet accompanying the metals and welding rods should be examined to determine the required type of respirator cartridge.

Paints and Thinners

Workers tend to be particularly casual and unjustifiably lax when dealing with paints and thinners. Nevertheless, paints and thinners are not benign, and certain precautions are important to prevent long-term health problems.

Personal Protective Equipment

The use of cloth coveralls and a lightweight cloth cap is strongly recommended. Paints tend to splatter on even the most experienced painters. Cloth gloves are recommended for the same reason. When painting above shoulder height, and particularly when painting overhead, eye goggles are required.

Brush and roller cleaning should be done in a well-ventilated area using rubber-lined or plastic gloves. Breathing protection should not be required if ventilation is good, but when painting in confined or poorly ventilated areas, an air purifying respirator (APR) is strongly recommended. The Material Safety Data Sheet for the paint being used will identify the specific respirator to use.

Safe Work Practices

As with other flammable materials, the areas in which paints and thinners are being used should be posted as "no smoking" areas, and the ban should be strictly enforced.

Phenolic Compounds

Phenolic compounds are typically painted or sprayed onto exterior surfaces or squeezed into cracks from caulking guns. Those compounds that are painted or sprayed are the most troublesome from a personal safety standpoint.

Personal Protective Equipment

Cloth coveralls, gloves, and hats should be worn at all times when dealing with these products. If spray application is being conducted, or if painting is being done at heights greater than shoulder level, eye goggles are essential. Since most applications are made to the outside of structures, ventilation is generally good and breathing protection is not required. If application is being made below ground level, in a trench, inside an enclosed space, or in any other poorly-ventilated area, an air purifying respirator (APR) is required. The Material Safety Data Sheet for the product being used will dictate the specific type of respirator to use.

Safe Work Practices

Due to the highly volatile nature of phenolic compounds, smoking must be prohibited by those working with these materials.

Oils and Lubricants

Oils and lubricants pose no significant health hazard to workers under normal circumstances. Rubberized gloves are suggested for those using these materials, and eye goggles are recommended if spray lubricants are being used.

Safe Work Practices

Sprays and lubricant vapors are flammable. No-smoking bans must be imposed on those using such materials. Spray lubricants should not be used near open flames or sparks.

Wallboard

The dust generated from the sanding of wallboard jointing compound contains caustic alkalis which can damage eyes, lungs, and mucous membranes. The compound itself will dry the skin during prolonged contact.

Personal Protective Equipment

When working with wallboard, workers should wear cloth gloves, eye goggles, and dust masks. Eye goggles and dust masks are required when applying or sanding joint compound.

Wood Products

Plywood and lumber products present a danger from splinters which, if ignored or improperly treated, will lead to infection. Of greater concern, however, is the sawdust associated with cutting and sanding wood. All wood dusts are hazardous to man. Certain wood dusts are toxic, although these are generally from the so-called "exotic" woods rarely found on construction sites. Even pine and fir, however, generate sawdust which can be carcinogenic when inhaled. Futhermore, dust from wood that has been treated with a chemical preservative is also toxic.

Personal Protective Equipment

Dust masks are required for all wood sawing and sanding operations. In addition, the high potential for chips being thrown about by these operations dictates that eye goggles also be worn when working with wood.

Chapter 6

Spill Prevention and Cleanup

An effective spill prevention program and an adequate spill cleanup plan are the marks of a safety-conscious contractor. Spills are inevitable on a construction site. How they are handled will directly affect the health and safety of the site workers.

This section will address spill prevention and containment techniques. (See Chapter 1 for a discussion of the notification requirements in the event of a spill. Uncontrolled spill cleanup is discussed in Part II of this book.) Construction details will be shown for the various devices and techniques recommended and outlined. Bear in mind, the costs of such prevention measures are always far less than those of the cure. Uncontained spills pose a very serious health hazard and a substantial fiscal liability.

Acids

Acids used for cleaning building surfaces are diluted, and spill prevention is not possible when using the dilute forms on exterior surfaces. The damage can be minimized, however, by using the following procedure and materials. For the worker's safety, rubber gloves should always be worn when working with acids. The next problem is to protect the soil and groundwater from the corrosive effects of acid. Since it is generally impractical to prevent the acids from reaching the soils, it is important to neutralize the acids that do spill. The easiest and least costly method is with the "universal base," lime.

Figure 6.1 shows a method that involves a shallow trench at the base of vertical walls that are being washed. The trench is filled with crushed lime and the wash water is allowed to percolate through the lime into the soil. This is an on-site treatment system which is low-cost and effective, and the residual lime is innocuous in the soil. Note that broadcasting lime outside the trench is a useful way to neutralize splashed liquids or spray mists. Figure 6.2 shows an adaptation of the vertical wall trench method to horizontal surfaces, with the same effects.

Dilute solutions usually start as concentrates which are mixed on-site with clean water. The mixing stage is a likely opportunity for a serious spill. Mixing of acids into water (never water into acids) should be done inside a containment box of adequate size to contain the full contents of the concentrate container. Figure 6.3 shows a low-cost wood and plastic containment system of adequate size to contain a five-gallon container of concentrate. Note that the bottom of this device is also coated with lime and that additional lime is available nearby to use in the event of a large spill.

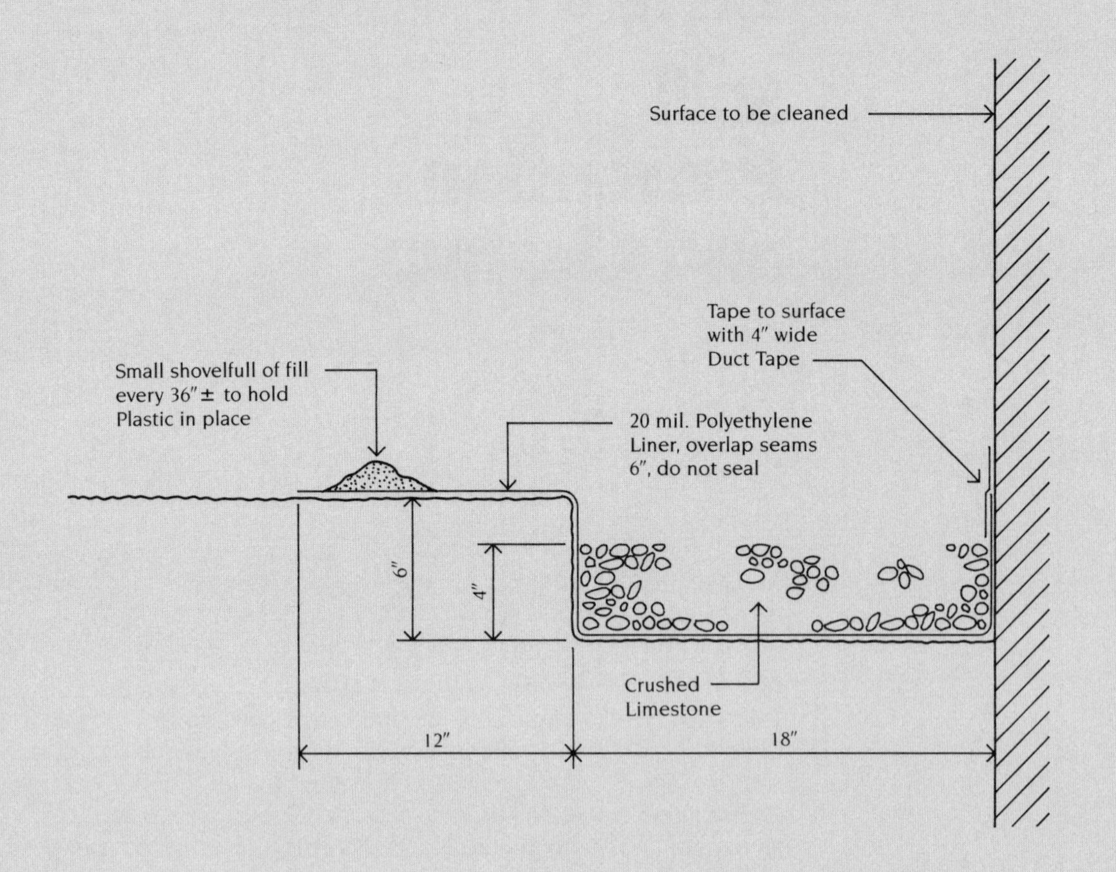

Note:
After Surface is cleaned, remaining Limestone may be reused or left in Trench. Plastic sheet may be reused or disposed of as non-hazardous waste. Wash water which has percolated through Limestone may be allowed to exfiltrate to surface soils without risk.

Bill of Materials Per Lineal Foot of Trench
0.75 cu. ft. Excavation and Backfill
12" × 42" Wide 20 mil. Polyethylene Sheet
0.5 cu. ft. Crushed Limestone
12" × 4" Wide Duct Tape

Acid Wash Containment Trench
for Vertical Work

Figure 6.1

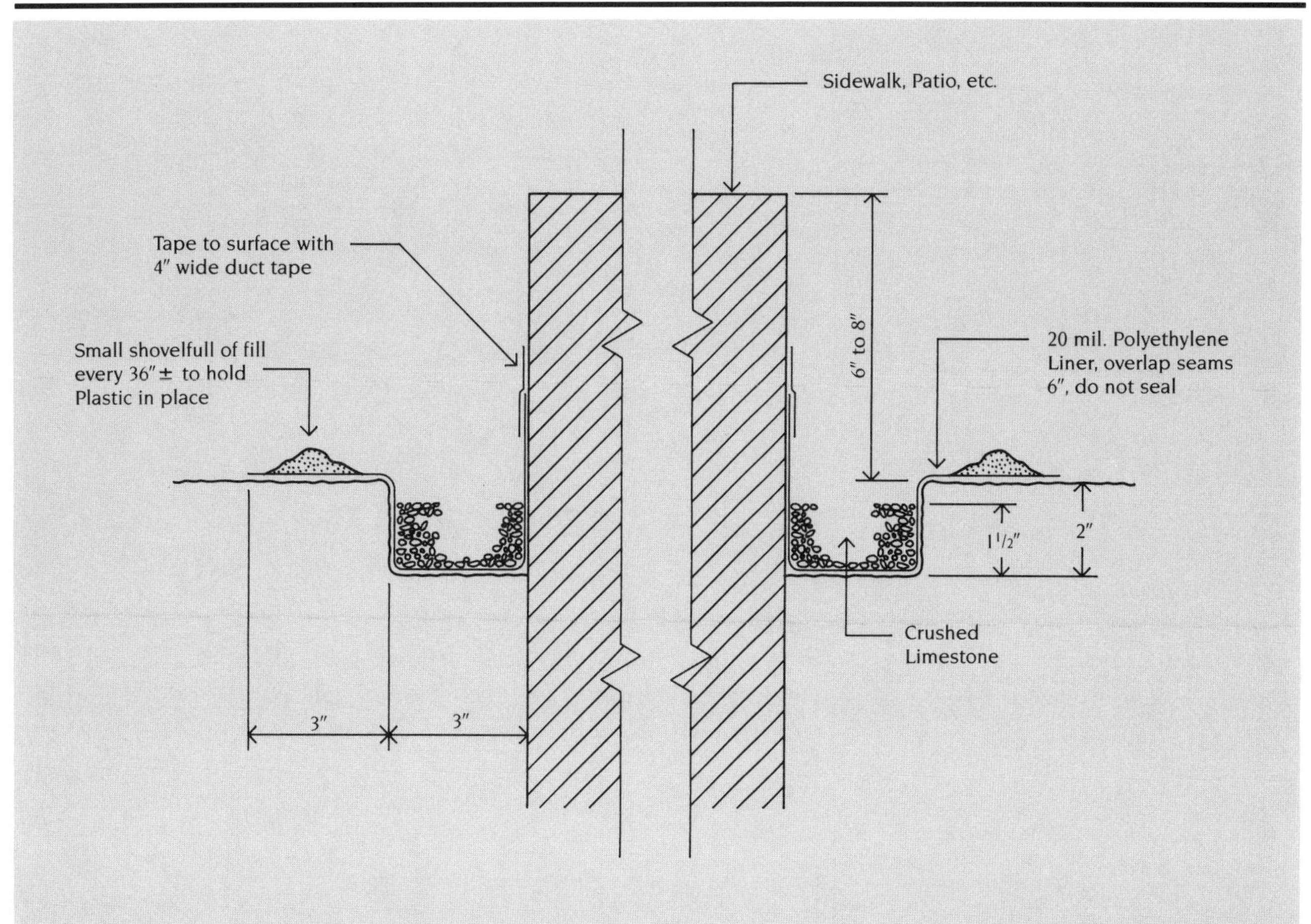

Note:
After Surface is cleaned, remaining Limestone may be reused or left in Trench. Plastic sheet may be reused or disposed of as non-hazardous waste. Wash water, with Acid, which has percolated through Limestone, may be allowed to exfiltrate to surface soils without risk.

Bill of Materials Per Lineal Foot of Trench
0.5 cu. ft. Excavation and Backfill
12" × 12" Wide 20 mil. Polyethylene Sheet
0.03 cu. ft. Crushed Limestone
12" × 4" Wide Duct Tape

Acid Wash Containment Trench
for Horizontal Work

Figure 6.2

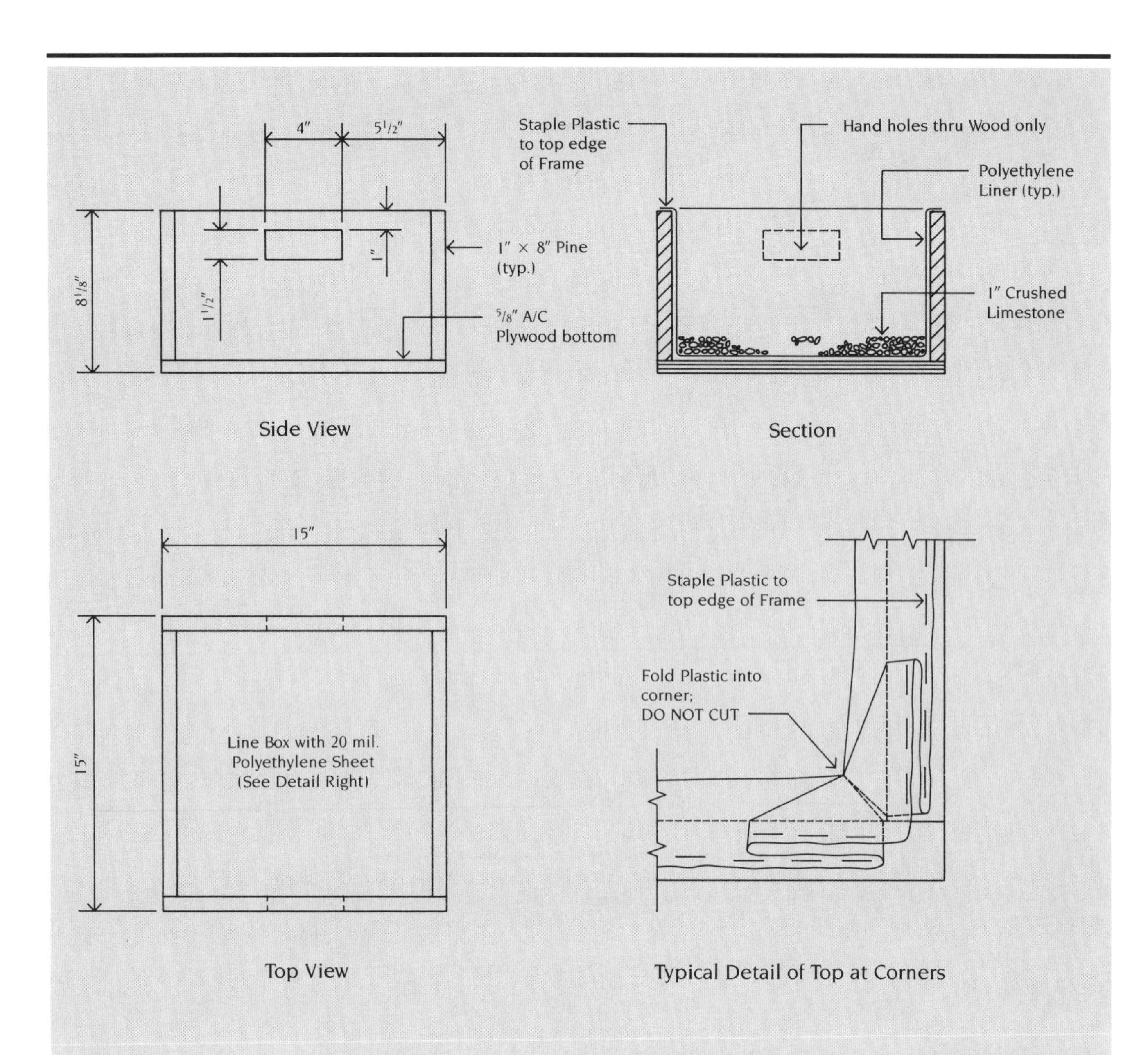

Bill of Materials

2 @ 15″ × 1″ × 8″ Pine
2 @ 13 3/4″ × 1″ × 8″ Pine
1 @ 15″ × 15″ × 5/8″ A/C Plywood
1/2 lb. 6d Galvanized Nails
1 @ 32″ × 32″ 20 mil. Polyethylene Sheet
10 lbs. Crushed Limestone

Notes:

Line inside of Box with Polyethylene Sheet. Then place 1″ Crushed Limestone over Liner.
Secure all joints with 6d Galv. Nails.

Spill Containment Box

Figure 6.3

Once the fizzing has stopped, the acid has become neutralized and the remaining lime is safe to dispose on the ground. It is recommended that the containment box be flooded with water to wash the lime out after a spill, and that a new absorption layer be put into the dike before the next batching operation is started.

Battery acids are usually transferred from container to battery inside a workshop. Spilling while pouring from the shipping container to the transfer bottle is best controlled by using a pouring box as shown in Figure 6.4. It is recommended that acids not be added to batteries while the battery is still in the vehicle or equipment. Move the battery to a pouring box for filling.

Acids dripping on the outside of a container or battery, or onto unprotected surfaces, should be immediately wiped with a clean rag dipped in a baking soda-and-water or lime-and-water paste to neutralize the acid. The used rag can be rinsed in clean water and reused. Dumping dry baking soda into a surface spill is also an effective neutralizing action.

Alkalis

The types of alkalis used on construction sites tend to be caustic when in direct contact with people, and corrosive in contact with metals. They are not particularly dangerous in soil, however. Spills are usually caused by bags of dry powder breaking open. The alkali materials are an ingredient in cement mortars and, therefore, harden with the first rain rather than leaching into groundwaters. Cleanup is accomplished by shoveling the dry powder into an appropriate container or disposing of solidified mortars to landfills.

Cleansers and cleaners containing only alkalis are diluted in use. Spills can be cleaned up with a generous amount of water to further dilute the materials.

Highly caustic drilling fluids and other waste liquids should be handled and disposed in accordance with the instructions given on the Material Safety Data Sheets provided with the material in question. Surface spills of these liquids should be absorbed with a dry absorbent. The wet absorbent/soil mixture should be disposed as a hazardous waste.

Chlorinated Solvents

Chlorinated solvent spills pose one of the most serious risks to the environment of the substances used for construction activities. The chlorine element appears to be the principal facilitator in all sorts of carcinogenic reactions, and the carrier solvents tend to move quite easily through most soil matrices to the groundwater table. Even a one-gallon spill of these substances has the potential to destroy up to one million gallons of drinking water.

Solvents of any kind, but particularly chlorinated solvents, should only be used in non-leaking dip vats or over concrete pads or metal pans on which dry absorbent has been spread to catch drips and spills. Spray cans should be used only in very well-ventilated areas over concrete pads or metal pans. Note that the grease and dirt removed by the solvent is itself a hazardous waste which requires special disposal considerations. If left on the ground, the dirt will also leach chlorinated solvents, plus other hazardous constituents dissolved from the dirt, into the groundwater. Where the likelihood of a spill is high, such as at the end of a pump spout connected to a solvent drum or under a spigot tapped into the end of a horizontal drum, special care is needed to avoid spills.

Figure 6.5 depicts a typical drain pan which can be placed around a 35-gallon or 55-gallon drum to catch pump drips. Note that the drum is centered in the pan and that dry absorbent is placed around the drum.

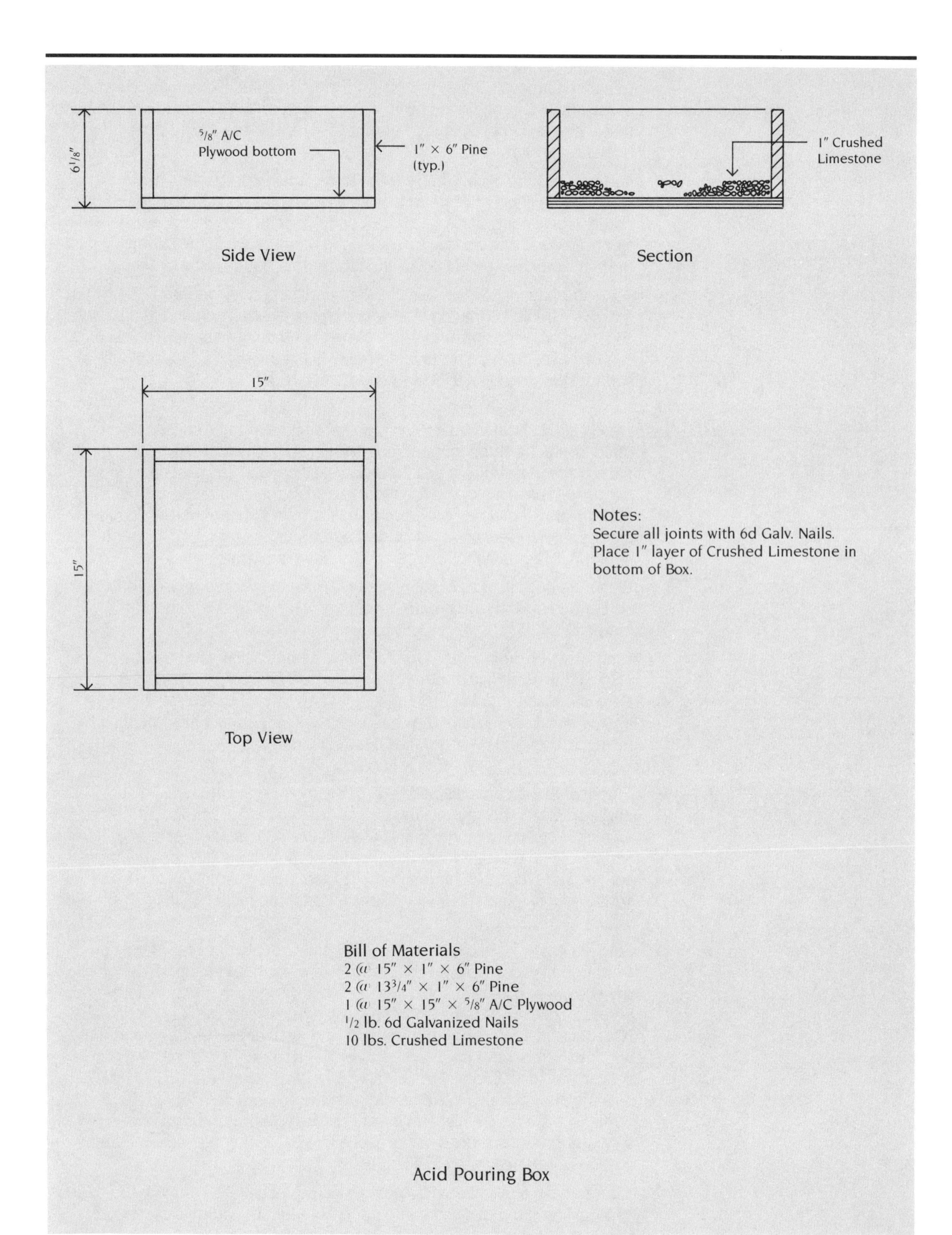
6 1/8"
5/8" A/C
Plywood bottom
1" × 6" Pine
(typ.)
Side View
1" Crushed
Limestone
Section
15"
15"
Top View
Notes:
Secure all joints with 6d Galv. Nails.
Place 1" layer of Crushed Limestone in
bottom of Box.
Bill of Materials
2 @ 15" × 1" × 6" Pine
2 @ 13 3/4" × 1" × 6" Pine
1 @ 15" × 15" × 5/8" A/C Plywood
1/2 lb. 6d Galvanized Nails
10 lbs. Crushed Limestone
Acid Pouring Box

Figure 6.4

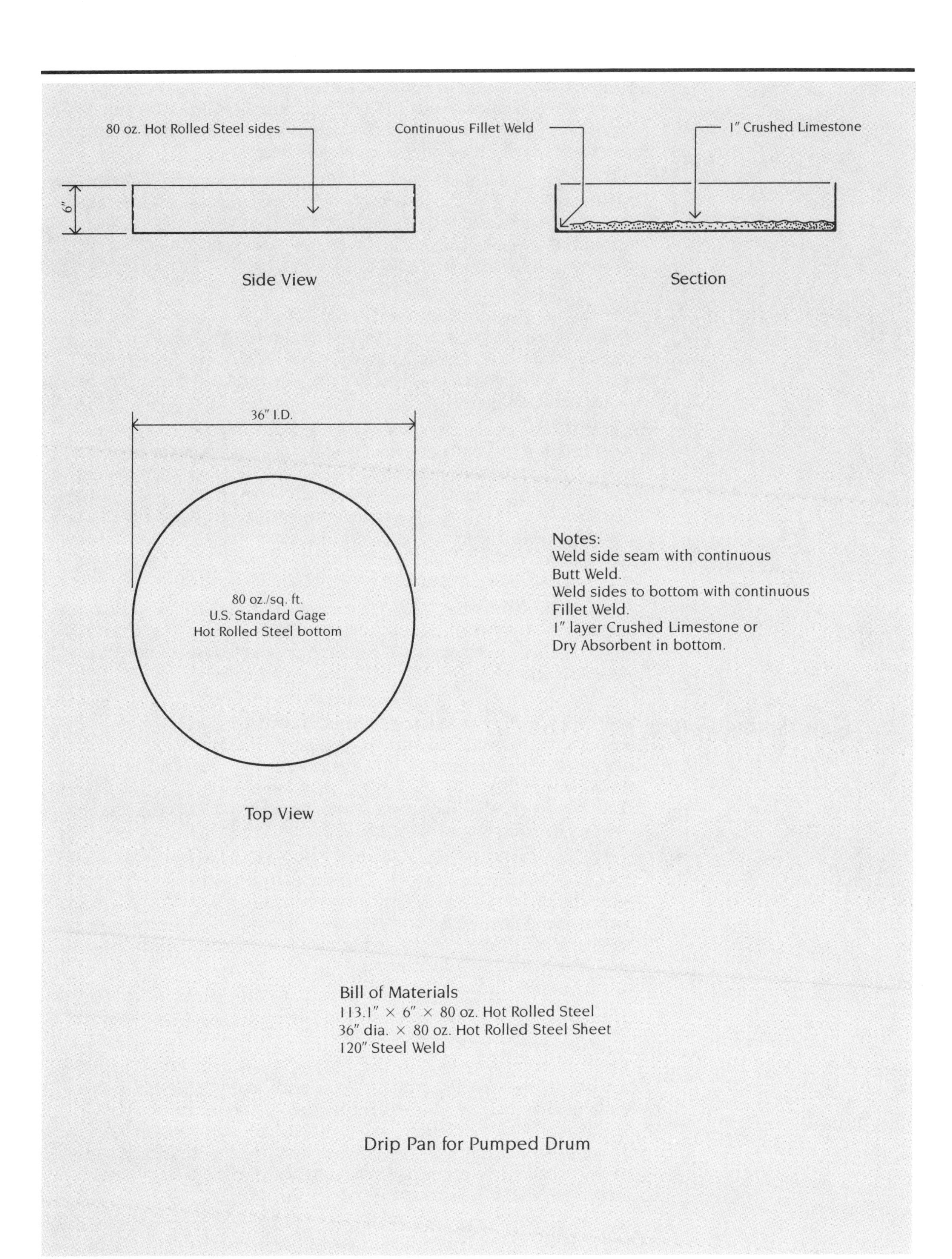
80 oz. Hot Rolled Steel sides
Continuous Fillet Weld
1″ Crushed Limestone
6″
Side View
Section
36″ I.D.
80 oz./sq. ft.
U.S. Standard Gage
Hot Rolled Steel bottom
Top View
Notes:
Weld side seam with continuous
Butt Weld.
Weld sides to bottom with continuous
Fillet Weld.
1″ layer Crushed Limestone or
Dry Absorbent in bottom.
Bill of Materials
113.1″ × 6″ × 80 oz. Hot Rolled Steel
36″ dia. × 80 oz. Hot Rolled Steel Sheet
120″ Steel Weld
Drip Pan for Pumped Drum

Figure 6.5

Figure 6.6 shows a standard drip pan for use under horizontally-mounted drums with end-tapped spigots. Note the dry absorbent inside the pan. While Figure 6.6 shows a drip pan for a single drum, a similar pan, of appropriate dimensions, can be used under a multi-drum rack.

Dry absorbent contaminated with chlorinated solvent must be disposed as a hazardous waste. *Non*-chlorinated solvents can be incinerated, but it is unsafe to incinerate any solvents containing chlorine. Moreover, normal landfills do not provide adequate groundwater protection to allow disposal of chlorinated solvents.

Chlorine

Chlorine gas is deadly. In the event of gas leakage, *immediately* evacuate the area of contamination and notify the fire department for assistance. If shutting off the flow of gas is feasible, it should only be done by a person wearing a self-contained breathing apparatus, having been properly trained in the use of this device.

Liquid chlorine, or chlorine solutions, are less dangerous, but very corrosive, nevertheless. A spill outdoors should be contained with sand or dry absorbent, and the contaminated soil should be placed on a plastic sheet and shielded from precipitation, but in a very well-ventilated area. Chlorine is highly volatile and will rapidly evaporate from the contaminated soil, at which point the soil can be reused on the site. Isolation of contaminated soil prior to dissipation of the chlorine is essential to avoid washing of the chlorine into the groundwater, from which it will not evaporate easily.

Care must be taken when disinfecting water lines to avoid improper disposal of the spent disinfectant. *Dilute* solutions are best disposed to a sewer (not a storm drain). Concentrated solutions must be collected and disposed as a hazardous waste.

Compressed Gases

Leaking gas cylinders pose two problems in confined spaces: risk of suffocation to humans, and risk of explosion from the gas/air mixture. When a leak is detected, the area must be immediately evacuated and the local fire department notified. If it is possible to stem the leak easily, that should be done, but only by a worker using a self-contained breathing apparatus, who has been properly trained in the use of that device.

Any open flames near leaking gas should be immediately extinguished and operations causing sparks should be discontinued. It is generally recommended that electrical equipment near the gas, such as lights, should *not* be shut off, because the act of throwing the switch often sets off a small internal spark. However, electric motors and heaters in the vicinity should be shut off until the gas is dissipated.

Ventilation of the area, allowing for dissipation of the gas to the atmosphere is recommended for those gases likely to be found on a construction site.

Coolants

The most common cause of coolant spills on construction sites is improper draining of vehicle cooling systems. When radiators are allowed to drain to the ground, for example, coolants contaminate the soil at the point of discharge and are then washed or seep through the soil matrix to the groundwater. One gallon of coolant, when discharged to the ground, can easily contaminate over one half million gallons of groundwater to levels above safe drinking water standards.

Prevention of spills involves collecting used radiator fluids in buckets or pails and then transferring the liquid to a storage drum prior to ultimate disposal as a hazardous waste. It is important, however, to keep used coolants

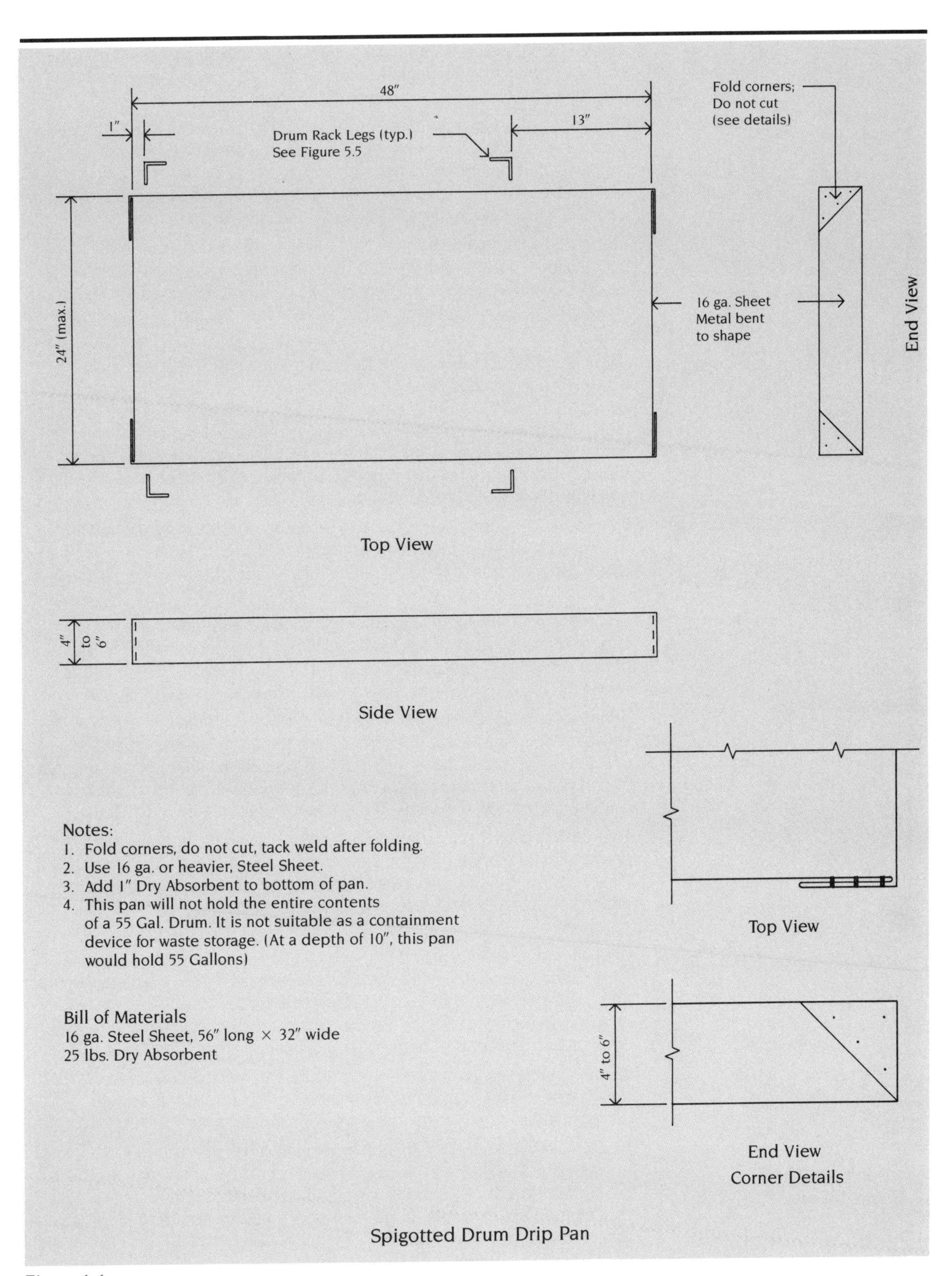
48"
1"
13"
Drum Rack Legs (typ.)
See Figure 5.5
Fold corners;
Do not cut
(see details)
24" (max.)
16 ga. Sheet
Metal bent
to shape
End View
Top View
4" to 6"
Side View
Notes:
1. Fold corners, do not cut, tack weld after folding.
2. Use 16 ga. or heavier, Steel Sheet.
3. Add 1" Dry Absorbent to bottom of pan.
4. This pan will not hold the entire contents of a 55 Gal. Drum. It is not suitable as a containment device for waste storage. (At a depth of 10", this pan would hold 55 Gallons)
Bill of Materials
16 ga. Steel Sheet, 56" long × 32" wide
25 lbs. Dry Absorbent
Top View
4" to 6"
End View
Corner Details
Spigotted Drum Drip Pan

Figure 6.6

separate from used oils and lubricants, because mixing makes the resultant waste far more difficult, and expensive, to dispose than either of the two original wastes.

Should a spill occur during refilling, use of a dry absorbent is recommended. Contaminated absorbent must be disposed as a solid hazardous waste, not merely landfilled, and any contaminated soil must be excavated quickly to avoid further leaching to groundwater. Excavated soils that are contaminated with coolants must be disposed as hazardous wastes.

Empty coolant containers should be flushed well with clear water before being disposed with normal refuse. The rinse water should be put into the cooling system since most coolants work best when diluted to a solution of about 50% water. Rinse water should not be disposed to the ground or sewer.

Flammable Solvents

Solvents of any kind should only be used in non-leaking dip vats or over concrete pads or metal pans on which dry absorbent has been spread to catch drips and spills. Spray cans should be used only in very well-ventilated areas over concrete pads or metal drip pans. Note that the grease and dirt removed by the solvent is itself a hazardous waste which requires special disposal consideration. If left exposed to moisture or precipitation, the dirt and grease will leach hazardous compounds to the groundwater.

Flammable solvents that do spill can be absorbed with a dry absorbent. The used absorbent must then be disposed as a hazardous waste. Rags or paper towels used to clean spills or to wipe surfaces dry must also be disposed as a hazardous waste.

Empty solvent containers may be safely disposed with normal refuse only after they have been air-dried for 36 to 48 hours in order to allow evaporation of the hazardous constituents. Empty containers should *not* be heated to speed the drying process, as this can lead to explosions and fires.

When solvents are shipped in large drums, there is a high probability of spillage when transferring the solvent from the drum to smaller transport containers. Figure 6.5 shows a typical drip pan which can be placed around a 35- or 55-gallon drum to catch pump drippings. Note that the drum is centered in the pan and that dry absorbent is placed around the drum.

Figure 6.6 depicts a standard drip pan for use under horizontally-mounted drums with end-tapped spigots. Note the dry absorbent inside the pan. While Figure 6.6 shows a drip pan for a single drum, a similar pan of appropriate dimensions can be used under a multi-drum rack.

Fuels—Gasoline/Diesel

Permanent fuel dispensing stations should be set on concrete pads with raised edges to contain spills. The pad should drain to an oil/water separator, and the drain should then be connected to a proper sewer (not a storm drain). Automatic fire suppression systems should be installed at all permanent dispensing stations to protect against fire and explosion.

Temporary fuel dispensing stations must be constructed in such a way that fuel cannot be spilled into the soil. Figure 6.7 shows a temporary fuel dispensing station suitable for use on a construction site.

When refueling vehicles on the job, care should be taken to avoid overfilling or splashing. If spillage to the ground does occur, it is necessary to excavate the contaminated soil as quickly as possible, isolate it from moisture and precipitation until it can be safely disposed, and dispose it as a hazardous waste.

Note that flushing soils contaminated with petroleum products merely

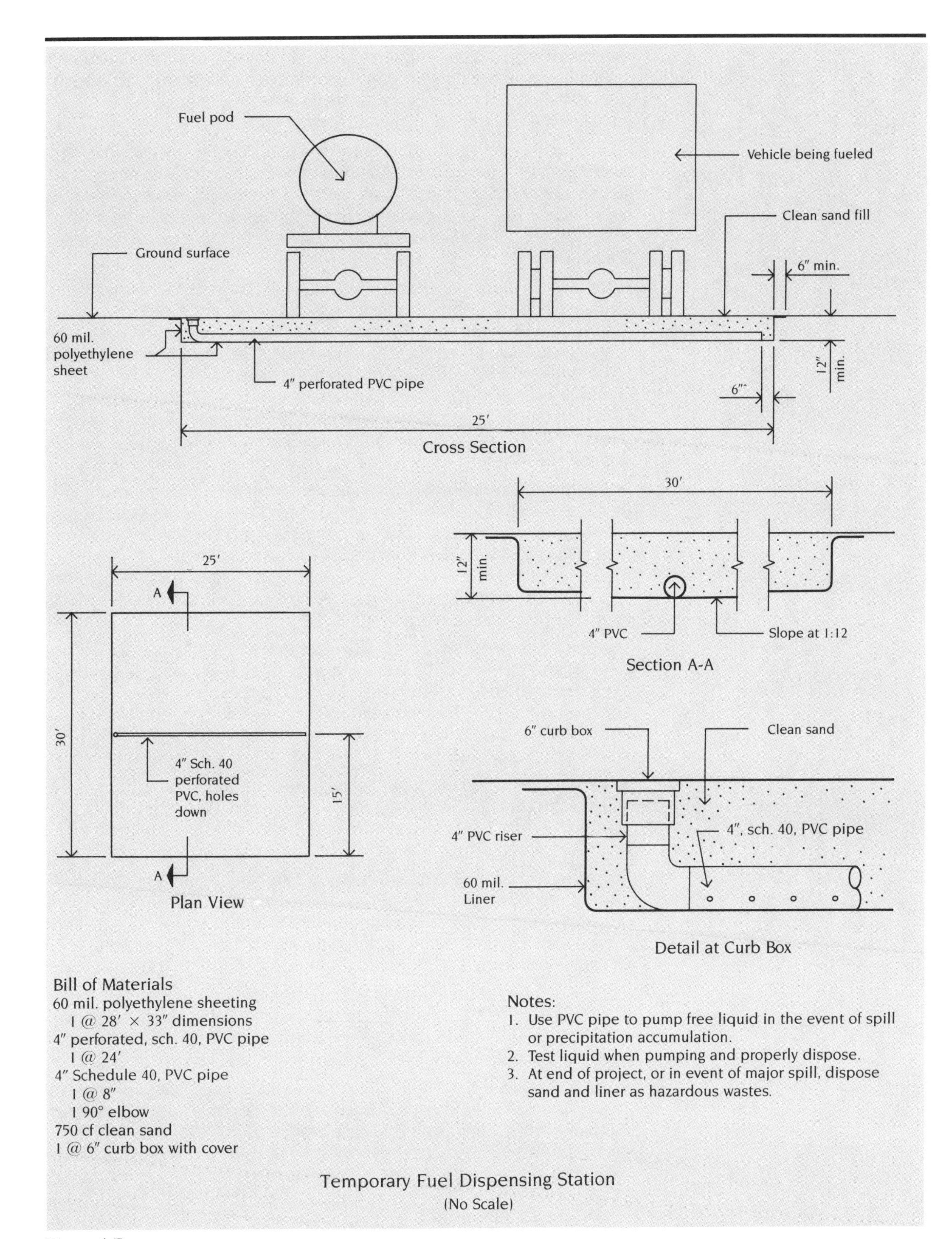
Fuel pod
Vehicle being fueled
Clean sand fill
Ground surface
6″ min.
60 mil. polyethylene sheet
4″ perforated PVC pipe
12″ min.
6″
25′
Cross Section
30′
12″ min.
4″ PVC
Slope at 1:12
Section A-A
25′
A
A
30′
4″ Sch. 40 perforated PVC, holes down
15′
Plan View
6″ curb box
Clean sand
4″ PVC riser
4″, sch. 40, PVC pipe
60 mil. Liner
Detail at Curb Box
Bill of Materials
60 mil. polyethylene sheeting
1 @ 28′ × 33″ dimensions
4″ perforated, sch. 40, PVC pipe
1 @ 24′
4″ Schedule 40, PVC pipe
1 @ 8″
1 90° elbow
750 cf clean sand
1 @ 6″ curb box with cover
Notes:
1. Use PVC pipe to pump free liquid in the event of spill or precipitation accumulation.
2. Test liquid when pumping and properly dispose.
3. At end of project, or in event of major spill, dispose sand and liner as hazardous wastes.
Temporary Fuel Dispensing Station
(No Scale)

Figure 6.7

exacerbates the problem. If solvents are put into a spill, the contamination moves through the soil to the groundwater more quickly than would otherwise be the case. That merely contaminates a larger volume of soil and expedites the contamination of underlying groundwater.

Paints and Thinners

Small paint spills (ten gallons or less), even on soil, generally do not pose a serious environmental threat. That is because they do not penetrate the soil matrix deeply, and they set up very quickly. The easiest way to deal with such a spill is to let it dry and then scrape up the soil/paint mixture for disposal with regular refuse.

Larger paint spills could penetrate soils to a dangerous depth before setting up and could leach solvents to the soil matrix as they dry. Therefore, larger spills should be mixed with a dry absorbent to contain them, and the contaminated soils should be excavated at once. The contaminated soils and absorbent should be isolated from the environment, such as by placing them on top of a sheet of heavy plastic and covering them with more plastic to prevent precipitation contact, until the paint has set up and dried. They can then be disposed as special wastes through a licensed hazardous waste disposal contractor.

Small indoor paint spills are best controlled with a dry absorbent which is then scraped up, isolated until dry, and disposed with normal refuse. Larger indoor spills are also best controlled using dry absorbent, but the used absorbent must be isolated from the environment until the paint dries. It must then be disposed as a special waste. Empty paint cans can be safely disposed with normal refuse only after being thoroughly air-dried for 36 to 48 hours.

Thinners pose a more serious threat in that they easily penetrate soil matrices and will seep directly to the groundwater. The larger threat appears to be from indiscriminate dumping of used solvents on job sites, rather than from accidental spills. The best way to avoid these occurrences is through proper *Right-to-Know* training and enforcement of written company policies prohibiting such practices.

To avoid accidental solvent spills, thinning of paints and cleaning of tools and brushes should only be done over a containment pan lined with dry absorbent. Figure 6.8 shows a containment pan suitable for small tools and brushes.

Spray-painting equipment must not be operated with thinners or solvents for cleaning purposes unless the spray is directed into a catchment device designed to either capture and reuse the solvent, or retain the contaminated materials for disposal. Contaminated thinners are flammable solvents which should be separately stored prior to disposal as hazardous wastes.

Thinners shipped in large drums present a high probability of spillage during transfer operations. Figure 6.5 illustrates a typical drip pan which can be placed around a 35- or 55-gallon drum to catch pump drippings. Note that the drum is centered in the pan, and that the pan is lined with dry absorbent.

Figure 6.6 depicts a standard drip pan for use under horizontally-mounted drums with end-tapped spigots. Note the dry absorbent inside the pan. While this figure shows a drip pan for a single drum, a similar pan of appropriate dimensions can be used under a multi-drum rack. Dry absorbents contaminated with thinners must be disposed as hazardous wastes.

Phenolic Compounds

Phenolic compounds tend to be slow to penetrate soil matrices, but they are also slow to harden and tend to leach phenols to the surrounding soils

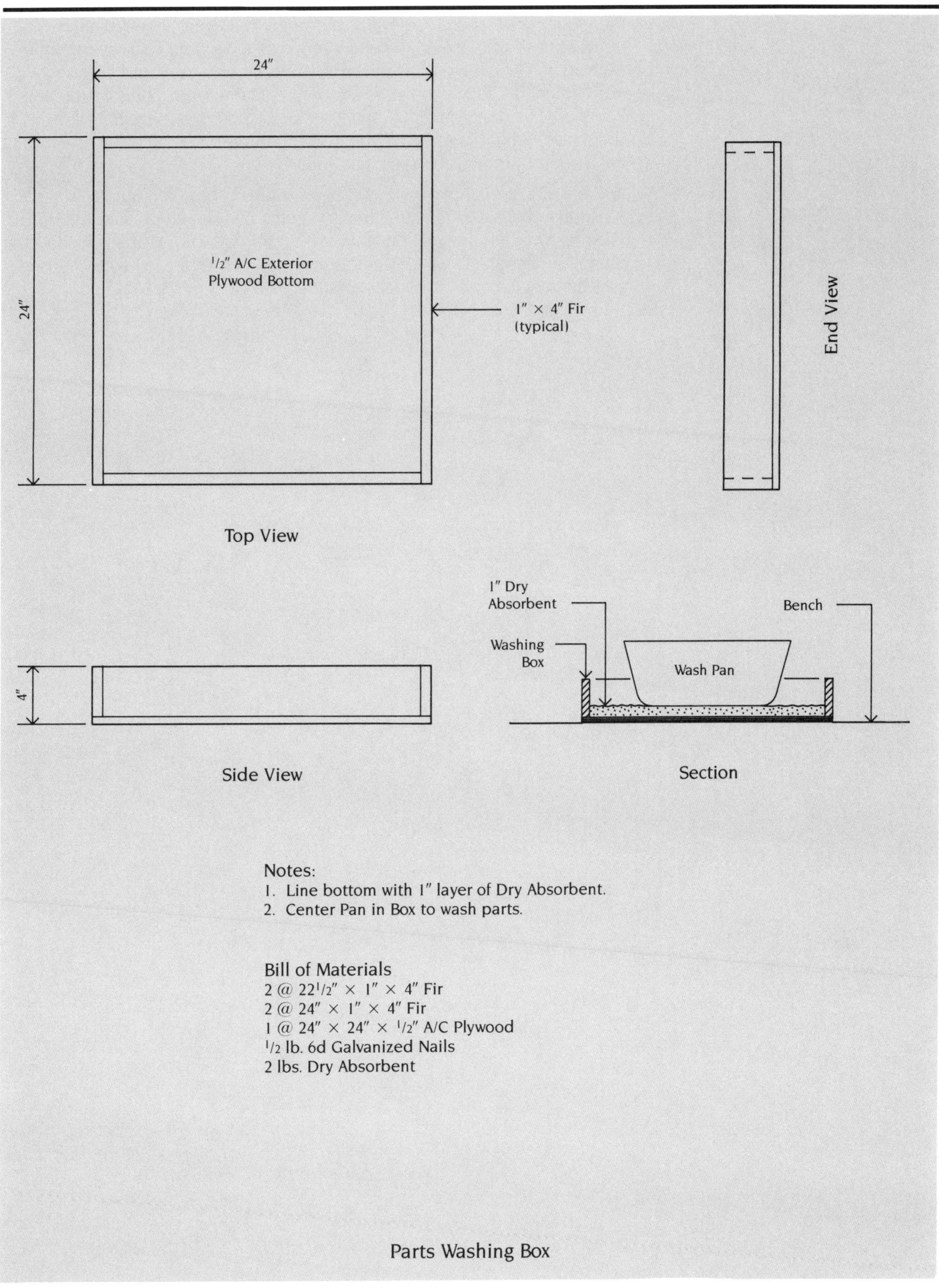
24"
24"
1/2" A/C Exterior
Plywood Bottom
1" × 4" Fir
(typical)
End View
Top View
1" Dry
Absorbent
Bench
Washing
Box
Wash Pan
4"
Side View
Section
Notes:
1. Line bottom with 1" layer of Dry Absorbent.
2. Center Pan in Box to wash parts.
Bill of Materials
2 @ 22 1/2" × 1" × 4" Fir
2 @ 24" × 1" × 4" Fir
1 @ 24" × 24" × 1/2" A/C Plywood
1/2 lb. 6d Galvanized Nails
2 lbs. Dry Absorbent
Parts Washing Box

Figure 6.8

during the hardening process. It is important, therefore, to mix sand or c absorbent with the spilled material immediately, in order to slow down the leaching process. The thickened mixture must then be isolated from the environment, as shown in Figure 6.9, until the compounds have set up and hardened. The dried material must then be disposed as a special waste. Alternatively, the thickened mixture can be drummed in 55-gallon containers and disposed immediately as a hazardous waste.

Indoor spills of phenolic compounds must also be covered quickly with sand or dry absorbent. The resulting material must be scraped up and, without delay, removed to an isolation station or drum due to the vaporization of the phenolic constituents which yield toxic fumes. Contaminated solvents used to clean final traces of phenolics off surfaces, as well as the rags used to apply the solvents, must be disposed as hazardous wastes.

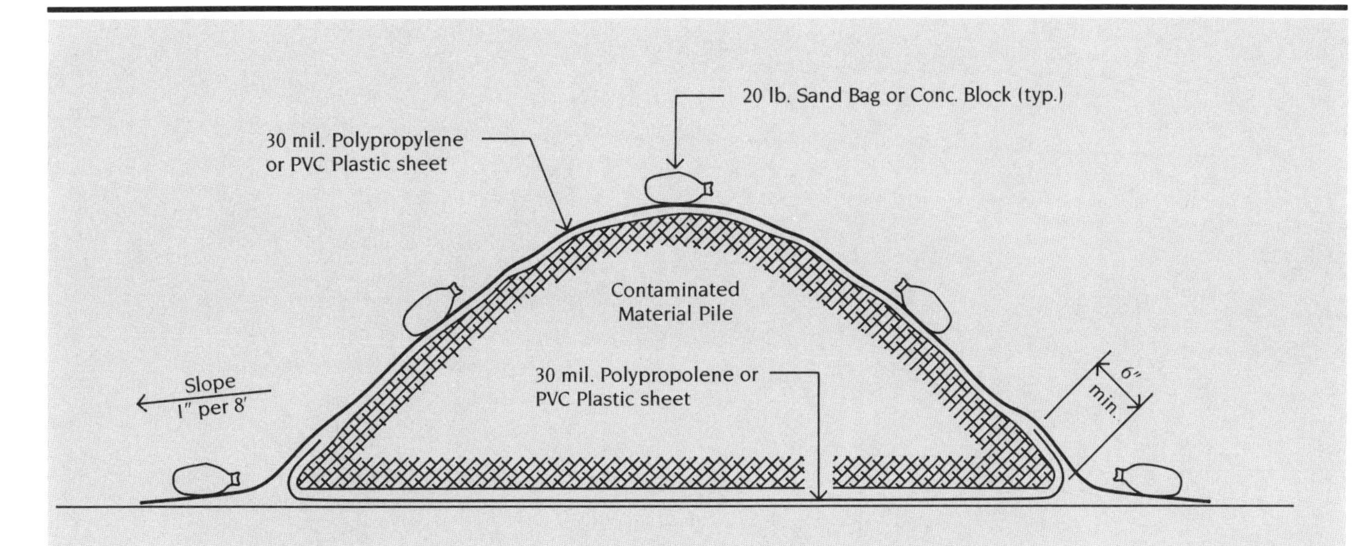

Section

Notes:

1. Place 30 mil. Plastic sheet on level surface.
2. Pile Waste Material in center of sheet.
3. Fold edges of Plastic sheet at least 6" onto outside of Pile.
4. Place separate 30 mil. Plastic sheet over Pile and secure against wind lifting with Sand Bags or Concrete Blocks.

Temporary Storage of Contaminated
Materials on Site

Figure 6.9

Chapter 7

Asbestos Handling

Asbestos is one of the most useful, yet dangerous substances on a construction site. Its properties of very low heat transfer and flexibility make it an excellent insulator and fire retardant. Its insidious carcinogenic nature makes it deadly.

The use of asbestos insulation for new construction is heavily regulated and generally banned for all practical purposes. There is a continued debate about the safety of using encapsulated asbestos fibers as a fire retardant. It is argued that spray application of asbestos, bound by an encapsulating resin, will not allow asbestos particles to become airborne and, therefore, poses no significant risk. A number of regulators and workers' groups are not yet convinced. Some states have banned the use of all insulations that contain asbestos.

It is generally recognized that asbestos is not hazardous by its mere presence. Only when the material is in a friable (easily crumbled) state, and then only when disturbed in such manner as to cause the friable particles to become airborne, does the material become hazardous.

Management protocols call for existing asbestos materials to be left undisturbed whenever possible. When disturbance is unavoidable, measures must be taken to minimize the potential for fibers to become airborne. Personal protective equipment must be used to prevent inhalation of particles that do become airborne, and proper cleanup procedures must be followed to ensure against the future possibility of floating residual particulates. This chapter presents the procedures to follow in order to accomplish these management objectives.

The presence of asbestos has become widespread through historical uses. It appears in a wide range of heat-, electrical-, and sound-insulating products, decorative facades, filter media, waterproofing, and other compounds. The Commonwealth of Massachusetts has identified about 50 common products that contain asbestos. Nearly all can be found on a construction site, particularly where demolition or renovation of older, existing buildings is in progress. Figure 7.1 lists the various products identified by the Commonwealth.

Storage

The most common asbestos-containing product currently in use on new construction is asbestos cement pipe. This type of pipe is used for water, drain, and sewer service in a variety of applications. The cement used to bond

Products Containing Asbestos		
Product	**Where Found**	**Use**
I. Friable Materials and Products		
a. Asbestos Cord	Electrical Installations and Equipment	Electrical Element Insulation
b. Asbestos Tape, Strip, and Tubing	Electrical Installations and Pipe Joints	Electrical Conductor Insulation, High Temperature Pipe Joint, Insulation Wrap
c. Fire-resistant Theater and Welding Curtains, Protective Clothing	Auditoriums, Stages, Metal Shops, High Temperature Occupations	Fire and Heat Barrier
d. Spray-applied or Troweled-on Insulation	Steel I-Beams and Decks, Concrete Ceilings and Walls, Hot Water Tanks, Pipe Elbows, Boiler Casing	Thermal and Acoustical Insulation, Decorative Coverings
e. Preformed Thermal Insulation	Boilers, Pipes, Hot Water Tanks	Thermal Insulation, Condensation Control
f. Artificial Snow	General Commercial Use	Decoration
g. Artificial Fireplaces, Logs for Gas-burning Fireplaces	General Commercial Use	Decoration
h. Corrugated Asbestos Paper	Hot Water, Steam Pipes	Thermal Insulation
II. Non-friable Materials and Products		
a. Acoustic Plaster	Corridors, Lunchrooms, Offices, Auditoriums, Music Rooms, Sound Control, and Projection Rooms	Acoustic Insulation
b. Asbestos Cement Products		
Transite-asbestos Panels	On Walls, Ceilings	Acoustical Purposes, Reinforcing Material
Storm Drainage Pipes	Drainage Systems	Reinforcing Material For Drain Pipe
Asbestos Stucco	Walls and Ceilings	Decorative Covering
Asbestos Siding Shingles	Usually on Exterior Walls	Decoration, Insulation and Protection of Building Structure
Roofing Shingles	Usually on Exterior Roof	Decoration, Insulation and Protection of Building Structure
Laboratory Table Tops	Laboratories	Heat and Chemical Resistance
Electrical Conduits	Electrical Installations	Insulating Material
Clapboard	Wood Frame Buildings Under Siding Material	Construction Material
Asbestos Ebony Products	Electrical Panels & Circuitry	Insulating Material

Figure 7.1

Products Containing Asbestos		
Product	**Where Found**	**Use**
c. Floor and Ceiling Tiles	—	Decorative, or Insulating Material
d. Clutch Facings, Drum Brake Linings, Disc Brake Pads and Brake Blocks	Automobiles, Railroad Cars, and Airplanes	Thermal Insulation, Reinforcing Material
e. Filtration Materials	Manufacturing and Food Processing Plants	Filters
f. Roofing Felts	Roofs of Buildings	Waterproofing and Weatherproofing
g. Roofing Asphalt	Roofs of Buildings	Waterproofing and Weatherproofing
h. Quality Vinyl Wallcoverings	—	Decorative Material
i. Gaskets	Varies	Heat Resistant Material
j. Millboard, Rollboard	Walls & Ceilings	Construction Material
k. Pipeline Wrap	Underground Oil and Gas Pipes	Corrosive Protection
l. Asbestos Insulation Board	Walls, Ceilings, Ducts, or Pipe Enclosures	Thermal and Fireproofing Barrier
m. Miscellaneous Compounds Containing Asbestos: Putties, Caulks, Paints, Sealants, Patching Tape Compounds, Tile Cement, Grout, Plaster, Oil/Gas Drilling Fluids	All Construction Occupations and Exploratory Drilling Occupations	Construction Materials and Lubricating Fluid
n. Shotgun Shell Base Wads	—	Thermal Insulation

From: "Asbestos Policy and Procedure Manual, Guidelines for Management and Maintenance Personnel," Massachusetts Division of Occupational Hygiene, 1001 Watertown Street, West Newton, MA 02165, pp. 7-3 through 7-5.

Figure 7.1 (*cont.*)

the asbestos in this pipe material prevents the asbestos fibers from becoming airborne under normal conditions. However, if the pipe is broken or cut, the exposed face often presents a high risk of exposure. Proper storage requirements include protection from accidental damage and ease of safe cleanup in the event of breakage.

Asbestos cement pipe and fittings should be stored outdoors on a level area that is not exposed to significant site traffic. Pipe should be chocked to prevent rolling, and fittings should be stored in wooden bins.

Other asbestos-bearing materials allowed for use in new construction will carry warning labels bearing proper handling instructions. In addition, MSDS forms provided with the product will list appropriate handling and storage procedures. In all cases, the manufacturer's instructions should be followed carefully.

Personal Protective Equipment (PPE)

The normal PPE used on construction sites is inadequate for working with asbestos. Normal dust filters are not capable of collecting the microscopic asbestos fibers which cause the greatest harm. Instead, high-efficiency particulate filters, designated "HEPA-filtered," are required. All HEPA-filtered respirators are reusable. (Single-use disposable filters do not meet the tightness of fit requirements for asbestos work.) Note that even reusable respirators require routine inspections, maintenance, and fit testing. The manufacturer's guidelines should be followed closely.

When using a half-face respirator, eye goggles are strongly recommended. In addition, because asbestos fibers can cling to clothing and may then be carried out of the work area where they can easily become airborne and contaminate other people, protective clothing is strongly recommended. Hair, clothes, and shoes all require protection.

Disposable suits, head coverings, and booties can be worn over regular clothing and then discarded before leaving the work area. Disposable clothing is only worn once, it is carefully removed in a decontamination area designated for the purpose, and is never worn outside the contaminated area. Once worn, the clothing must be carefully bagged and disposed as a hazardous waste. Each time a worker enters the work area, a new suit should be donned. Each time a worker leaves the work area, the old suit must be left behind. Note that a protective hood is worn outside the face mask straps.

Handling and Use

Whole libraries of books have been written on the subject of the handling and use of asbestos. Special training is required to perform this heavily regulated activity. An employee who has not been properly trained to do so should not be allowed to work with any asbestos material.

Those who do handle asbestos must wear proper and adequate personal protective equipment, operate in a manner to avoid damage to asbestos-bearing products, and minimize the potential for floating of particles in the air. In particular, workers should not cut, drill, or nail into asbestos. In addition, fixtures, wires, etc., should not be hung on or placed against asbestos products.

Renovation Work

The most difficult asbestos problem on a construction site is that of old, dry, friable asbestos hidden in the walls or fixtures of older buildings. By simply removing a wall to access the old material, a worker may face a serious exposure hazard.

The first thing to do when asbestos is discovered is to stop work. No time schedule is worth risking the health of a worker. Work should not resume until the extent of the hazard is known and proper management practices developed.

Second, call in an expert. Asbestos is not a matter to be taken lightly. The experienced worker will arrive equipped with proper face masks, protection clothing, HEPA-filtered vacuum equipment, etc.

If the work area is dusty or disturbed, and if removal is going to be necessary, a water spray on the asbestos and debris will dramatically reduce the risk of exposure. However, if the existing asbestos is to be encapsulated, spraying the dry asbestos with water will complicate the process. In that case, only the debris pile and general work areas should be wetted.

Special rules and regulations govern the disposal of asbestos waste. In general, they require double bagging in 6 mil, or thicker, plastic bags, sealed with duct tape and disposed to a licensed asbestos disposal contractor.

Cleanup

Cleanup after use or removal of asbestos requires vacuuming of carpets, rugs, furniture, and wall coverings with a HEPA-filtered vacuum. A conventional vacuum cleaner must never be used to clean up asbestos debris. Floor areas should be wet mopped. All horizontal surfaces should be wiped with a damp rag. Mopheads, rags, etc., should be disposed in a pre-labeled six-to-ten mil polyethylene bag, sealed with duct tape. A second bag should be sealed over the first in the same manner. Wash water can be disposed to a sink, but the buckets and sink must be thoroughly cleaned afterwards. Chapter 13 addresses asbestos waste disposal in more detail.

HAZARDOUS MATERIAL AND HAZARDOUS WASTE

Part II

Chapter 8

The Resource Conservation and Recovery Act: An Overview

The Resource Conservation and Recovery Act, more commonly known by its acronym RCRA, was enacted by Congress in 1976 to convey amendments to its predecessor, the Solid Waste Disposal Act of 1965. The Solid Waste Disposal Act of 1965 had been enforced by the Department of Health, Education and Welfare (now known as the Department of Health and Human Services). The emphasis of the Solid Waste Disposal Act was the regulation of refuse and, in particular, restrictions on open dumping.

In 1970, with heightened public awareness of the need for environmental protection and the formation of the Environmental Protection Agency (EPA), Congress amended the Solid Waste Disposal Act and renamed it the *Resource Conservation and Recovery Act*. The amended statute mandated extensive research and investigation into alternative hazardous waste management techniques, and provided appropriations for the collection and recycling of refuse materials. The newly formed EPA was given authority to administer the Resource Recovery Act.

The Solid Waste Disposal Act of 1965, together with its 1970 Resource Recovery Act amendments, were, for the most part, ineffective in addressing the nation's waste management problems. It was for this reason that Congress superseded the Solid Waste Disposal Act and its amendments with the text of RCRA. RCRA was enacted as, and continues to be, the primary law governing hazardous waste management.

As the time for reauthorization drew near, RCRA was criticized for its inadequacies, and a more stringent regulation was demanded. Thus, Congress passed the Hazardous and Solid Waste Amendments of 1984, not only as a reauthorization of RCRA, but as a more forceful version of RCRA. The amendments clarified ambiguities, closed certain loopholes in the original act, and also imposed more stringent requirements on several aspects of hazardous waste management. In addition, they enlarged the scope of RCRA itself and of RCRA's regulatory programs.

Hazardous Waste Management

Subchapter III of RCRA, entitled "Hazardous Waste Management," is one of the most significant sections of the Act with respect to the construction industry. This subchapter establishes what has been referred to as a "cradle-to-grave" regulatory scheme for the management of hazardous wastes. Its provisions impact the generation, transportation, storage, treatment, and disposal of hazardous waste.

Is a Contractor a Generator?

Deciphering federal regulations is not always an easy task. It may not be clear to the reader whether or not a particular regulation applies to him or her. Not all contractor operations generate hazardous wastes. It is useful, then, to consider a simple method for determining whether one is or is not a generator of hazardous waste.

The first question to answer is whether one generates a *solid waste*. Under RCRA, only those wastes that meet the definition of solid waste provided in the RCRA regulations can be a hazardous waste. (*Warning*: the definition of *solid waste* under RCRA includes certain gases and liquids. It is a much broader definition than that normally used to describe a solid waste.)

RCRA classifies all materials according to three basic categories:

1. Garbage, refuse, or sludge
2. Solid, liquid, semi-solid, or contained gaseous material
3. Something else

All materials in the first category are solid wastes. None of the materials in the third category are solid wastes. Materials in the second category are solid wastes unless they fall into one of five exclusions specified.

Materials in category 2 are excluded from the definition of solid waste if they are:

1. Domestic sewage
2. A point discharge source regulated under the Clean Water Act
3. An irrigation return flow
4. A special nuclear or by-product material regulated by the Atomic Energy Commission
5. An insitu mining waste

It is unlikely that any waste generated at a construction site will meet one of the exclusions, although there is a slight possibility that exclusions 1 and 3 may apply. Figure 8.1 is a logic diagram which may be used to help determine whether a waste meets the definition of solid waste for purposes of RCRA.

There are two criteria under which an RCRA solid waste may be defined as a hazardous waste. A waste is defined as hazardous if:

1. It is, or contains, a hazardous waste named in the RCRA regulations or listed by EPA (See Appendix A for these materials)
2. It exhibits any of four specific characteristics of hazardous waste defined in the RCRA regulations. These are: ignitability, corrosivity, reactivity, and Extraction Procedure (EP) Toxicity

A full definition of each of these characteristics can be found in Figure 8.2. Some exclusions are allowed for materials defined by the two cited criteria, but those exclusions would very rarely, if ever, apply to construction activities. Figure 8.3 is a logic diagram that may be used to help determine whether a solid waste, as defined by RCRA, is also a hazardous waste under RCRA.

If a contractor determines that he/she is a generator of hazardous wastes, then a determination must be made concerning the category of generator. Three categories of generator are recognized by EPA. They are: (1) Large Quantity Generator (LQG); (2) Small Quantity Generator (SQG); (3) Very Small Quantity Generator (VSQG).

The status is determined by the total quantity of hazardous waste generated in a one-month period. In determining the total quantities generated, waste oils are counted separately from other hazardous wastes.

Refer to Figure 1.1 in Chapter 1 for a definition of the accumulation time and volume limits applicable to the different categories of generator. Note, too, that some states regulate waste oils as hazardous wastes. Those states that

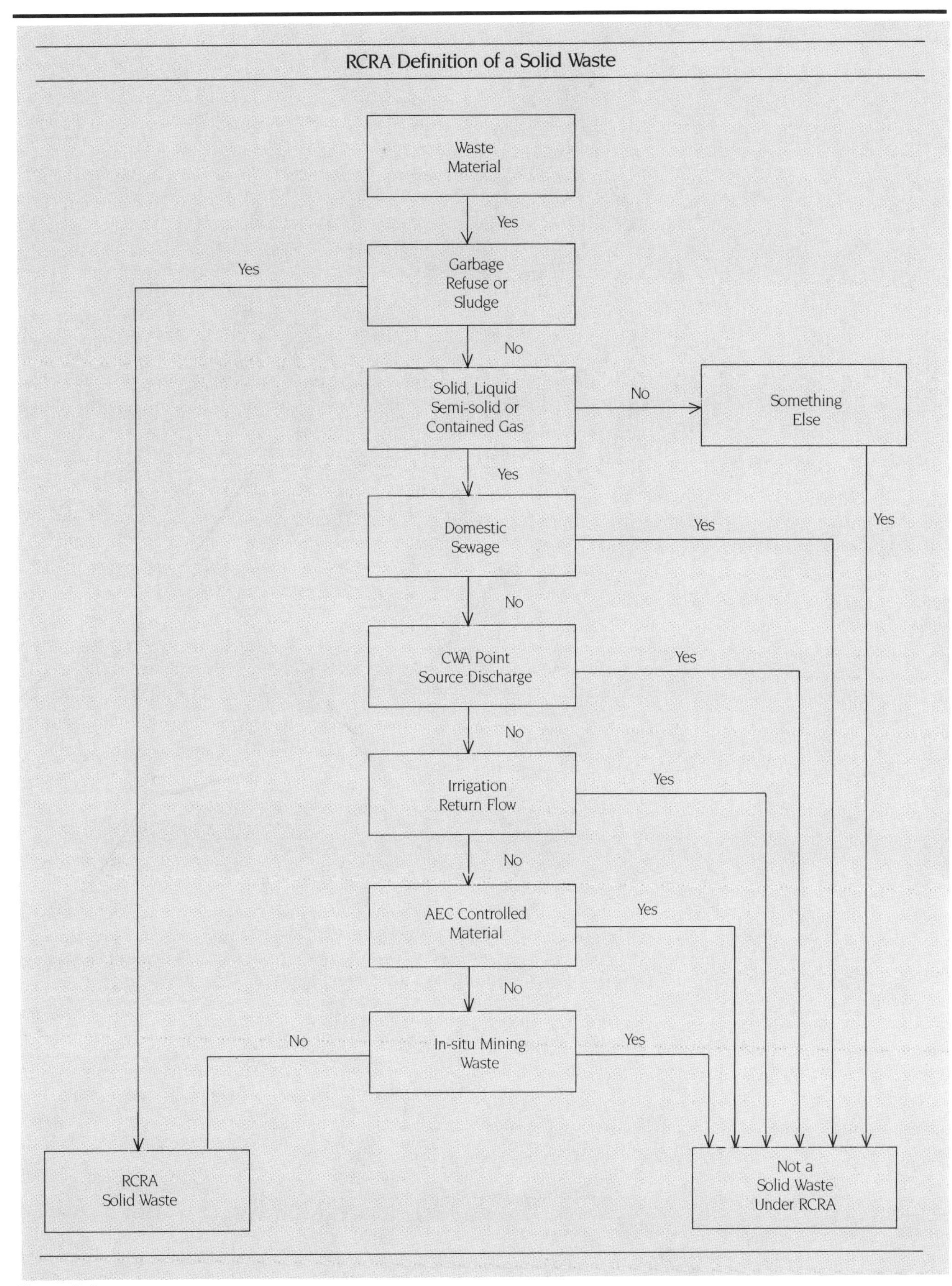
RCRA Definition of a Solid Waste
Waste Material
Yes
Garbage Refuse or Sludge
Yes
No
Solid, Liquid Semi-solid or Contained Gas
No
Something Else
Yes
Domestic Sewage
Yes
Yes
No
CWA Point Source Discharge
Yes
No
Irrigation Return Flow
Yes
No
AEC Controlled Material
Yes
No
In-situ Mining Waste
No
Yes
RCRA Solid Waste
Not a Solid Waste Under RCRA

Figure 8.1

Characteristics of Hazardous Wastes
RCRA Regulations

§ 261.21 Characteristic of Ignitability

(a) A solid waste exhibits the characteristic of ignitability if a representative sample of the waste has any of the following properties:
  (1) It is a liquid, other than an aqueous solution, containing less than 24 percent alcohol by volume and has a flash point less than 60°C (140°F), as determined by a Pensky-Martens Closed Cup Tester, using the test method specified in ASTM Standard D-93-79 or D-93-80 (incorporated by reference, see § 260.11), or a Setaflash Closed Cup Tester, using the test method specified in ASTM Standard D-3278-78 (incorporated by reference, see § 260.11), or as determined by an equivalent test method approved by the Administrator under procedures set forth in §§ 260.20 and 260.21.
  (2) It is not a liquid and is capable, under standard temperature and pressure, of causing fire through friction, absorption of moisture, or spontaneous chemical changes, and when ignited, burns so vigorously and persistently that it creates a hazard.
  (3) It is an ignitable compressed gas as defined in 49 CFR 173.300 and as determined by the test methods described in that regulation or equivalent test methods approved by the Administrator under §§ 260.20 and 260.21.
  (4) It is an oxidizer defined in 49 CFR 173.151.

(b) A solid waste that exhibits the characteristic of ignitability, but is not listed as a hazardous waste in Subpart D, has the EPA Hazardous Waste Number of D001.

[45 FR 33119, May 19, 1980, as amended at 46 FR 35247, July 7, 1981]

§ 261.22 Characteristic of Corrosivity

(a) A solid waste exhibits the characteristic of corrosivity if a representative sample of the waste has either of the following properties:
  (1) It is aqueous and has a pH less than or equal to 2 or greater than or equal to 12.5, as determined by a pH meter using either an EPA test method or an equivalent test method approved by the Administrator under the procedures set forth in §§ 260.20 and 260.21. The EPA test method for pH is specified as Method 5.2 in "Test Methods for the Evaluation of Solid Waste, Physical/Chemical Methods" (incorporated by reference, see § 260.11).
  (2) It is a liquid and corrodes steel (SAE 1020) at a rate greater than 6.35 mm (0.250 inch) per year at a test temperature of 55°C (130°F) as determined by the test method specified in NACE (National Association of Corrosion Engineers) Standard TM-01-69 as standardized in "Test Methods for the Evaluation of Solid

Figure 8.2

Characteristics of Hazardous Wastes
RCRA Regulations (continued)

Waste, Physical/Chemical Methods" (incorporated by reference, see § 260.11) or an equivalent test method approved by the Administrator under the procedures set forth in §§ 260.20 and 260.21.

(b) A solid waste that exhibits the characteristic of corrosivity, but is not listed as a hazardous waste in Subpart D, has the EPA Hazardous Waste Number of D002.

[45 FR 33119, May 19, 1980, as amended at 46 FR 35247, July 7, 1981]

§ 261.23 Characteristic of Reactivity

(a) A solid waste exhibits the characteristic of reactivity if a representative sample of the waste has *any* of the following properties:
  (1) It is normally unstable and readily undergoes violent change without detonating.
  (2) It reacts violently with water.
  (3) It forms potentially explosive mixtures with water.
  (4) When mixed with water, it generates toxic gases, vapors or fumes in a quantity sufficient to present a danger to human health or the environment.
  (5) It is a cyanide or sulfide bearing waste which, when exposed to pH conditions between 2 and 12.5, can generate toxic gases, vapors or fumes in a quantity sufficient to present a danger to human health or the environment.
  (6) It is capable of detonation or explosive reaction if it is subjected to a strong initiating source or if heated under confinement.
  (7) It is readily capable of detonation or explosive decomposition or reaction at standard temperature and pressure.
  (8) It is a forbidden explosive as defined in 49 CFR 173.51, or a Class A explosive as defined in 49 CFR 173.53 or a Class B explosive as defined in 49 CFR 173.88.

(b) A solid waste that exhibits the characteristic of reactivity, but is not listed as a hazardous waste in Subpart D, has the EPA Hazardous Waste Number of D003.

§ 261.24 Characteristic of EP Toxicity

(a) A solid waste exhibits the characteristic of EP toxicity if, using the test methods described in Appendix II or equivalent methods approved by the Administrator under the procedures set forth in §§ 260.20 and 260.21, the extract from a representative sample of the waste contains any of the contaminants listed in Table I at a concentration equal to or greater than the respective value given in that Table. Where the waste contains less than 0.5 percent filterable

Figure 8.2 (*cont.*)

Characteristics of Hazardous Wastes
RCRA Regulations (continued)

solids, the waste itself, after filtering, is considered to be the extract for the purposes of this section.

(b) A solid waste that exhibits the characteristic of EP toxicity, but is not listed as a hazardous waste in Subpart D, has the EPA Hazardous Waste Number specified in Table I which corresponds to the toxic contaminant causing it to be hazardous.

Table I–
Maximum Concentration of Contaminants for Characteristic of EP Toxicity

EPA Hazardous Waste Number	Contaminant	Maximum Concentration (milligrams per liter)
D004	Arsenic	5.0
D005	Barium	100.0
D006	Cadmium	1.0
D007	Chromium	5.0
D008	Lead	5.0
D009	Mercury	0.2
D010	Selenium	1.0
D011	Silver	5.0
D012	Endrin (1,2,3,4,10-hexachloro-1,7-epoxy-1,4,4a,5,6,7,8,8a,-octohydro-1,4- endo, endo-5,8-dimethano-napthalene.	0.02
D013	Lindane (1,2,3,4,5,6-hexa-chlorocyclohexane, gamma isomer.	0.4
D014	Methoxychlor (1,1,1-Trichloro-2,2-bis [p-methoxyphenyl]ethane).	10.0
D015	Toxaphene ($C_{10}H_{10}Cl_5$, Technical chlorinated camphene, 67-69 percent chlorine).	0.5
D016	2,4-D, (2,4-Dichlorophenoxy-acetic acid).	10.0
D017	2,4,5-TP Silvex (2,4,5-Trichlorophenoxypropionic acid).	1.0

Figure 8.2 (cont.)

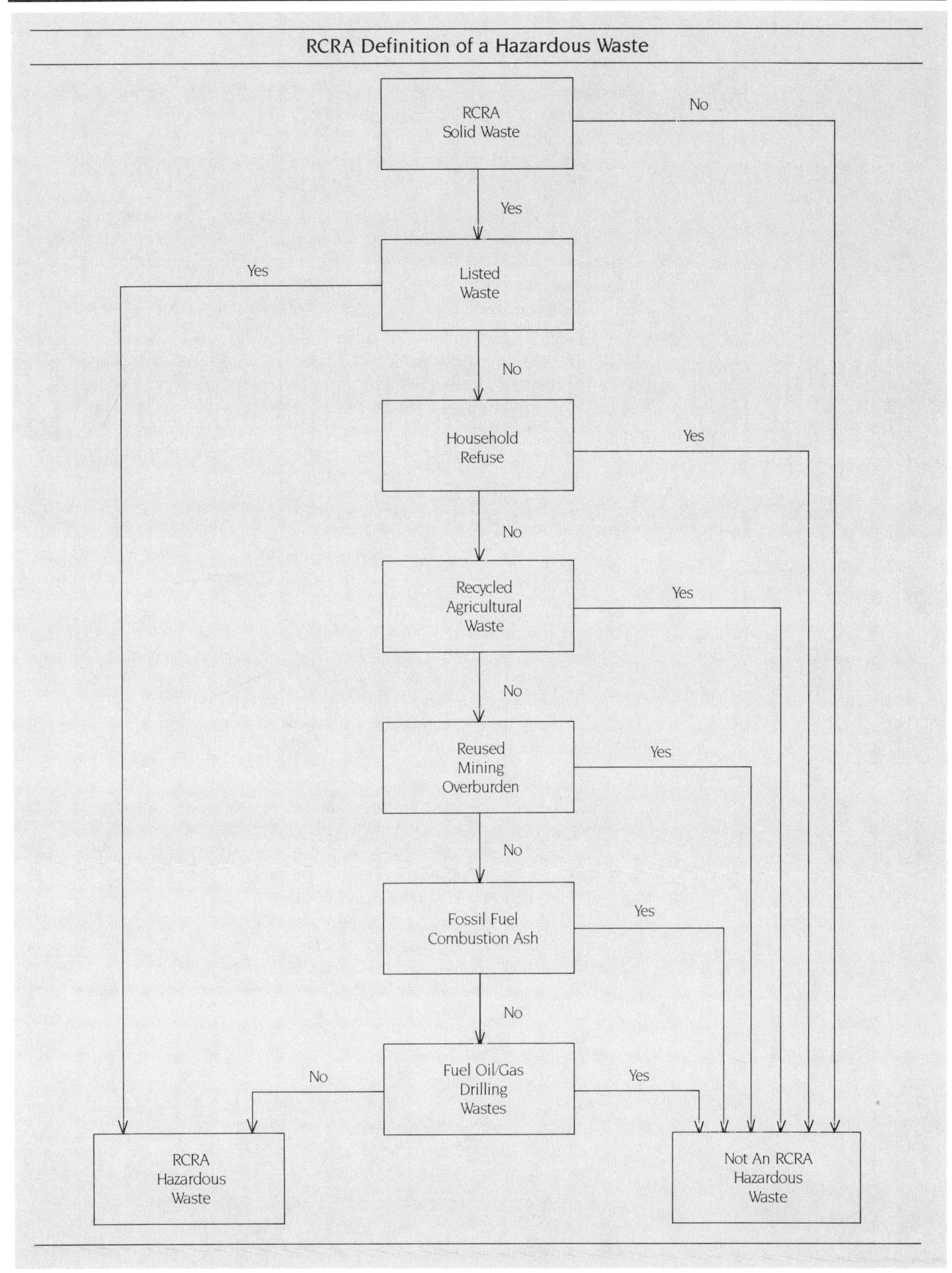
RCRA Definition of a Hazardous Waste
RCRA
Solid Waste
No
Yes
Listed
Waste
Yes
No
Household
Refuse
Yes
No
Recycled
Agricultural
Waste
Yes
No
Reused
Mining
Overburden
Yes
No
Fossil Fuel
Combustion Ash
Yes
No
Fuel Oil/Gas
Drilling
Wastes
No
Yes
RCRA
Hazardous
Waste
Not An RCRA
Hazardous
Waste

Figure 8.3

do regulate waste oil may also limit the total combined volume of hazardous waste and waste oil that may be accumulated on site.

Standards for Generators of Hazardous Waste

Contractors, as generators of hazardous waste, fall within the purview (limit of authority) of RCRA. Contractors are considered *generators* when the construction debris they produce is comprised of or contains hazardous waste constituents. Accordingly, contractors should be familiar with the statutory and regulatory provisions applicable to generators.

EPA has established specific requirements that apply to generators of hazardous wastes. These requirements pertain to record keeping, labeling, the use of containers, the furnishing of information to persons handling hazardous wastes, the use of a system, and the submission of reports to the EPA containing specific information on the hazardous wastes generated.

The regulations set forth a comprehensive hazardous waste tracking system. They force accurate identification of hazardous waste that is generated by requiring the use of EPA identification numbers and the completion of a standard form known as a *Manifest* whenever the generator transports hazardous waste or contracts others to remove hazardous waste. The regulations also contain pre-transport labeling, marking, placarding, accumulation, import, and export requirements.

For the purpose of the EPA regulations pertaining to generators of hazardous waste, it is important to note that a generator is defined as any person, by site, whose act or process produces hazardous waste identified or listed in the regulations.

EPA Identification Numbers

A primary requirement of the EPA regulations is that a generator obtain an EPA identification number before treating, storing or disposing hazardous waste; or transporting hazardous waste to treatment, storage, or disposal facilities. Identification numbers are obtained from the EPA by filing an application with the agency using EPA Form 8700-12. Figure 8.4 is a copy of EPA Form 8700-12.

The Manifest

The next major requirement is the Manifest. All generators who transport or contract for the transportation of hazardous waste for transportation for off-site treatment, storage, or disposal must prepare a Manifest. The Manifest is a standard EPA form, which lists the following.

- The generator's EPA identification number
- The generator's name and mailing address
- The generator's telephone number
- All transporters' names and EPA identification numbers
- The company name, site address, and EPA identification number for the facility designated to receive the waste
- The number and type of containers holding the waste
- The total quantity of waste
- The weight or volume of the waste
- The U.S. Department of Transportation description (including proper shipping name, hazard class, and identification number)
- Special handling instructions and other information
- The generator's written certification that the statements made are true

Figure 8.5 is a copy of a standard EPA manifest form. Note that if the state within which the waste is generated or the state within which the designated facility is located supplies a state manifest form, the generator must use that state's form.

Please print or type with ELITE type *(12 characters per inch)* in the unshaded areas only

Form Approved. OMB No. 2050-0028. Expires 9-30-88.
GSA No. 0246-EPA-OT

United States Environmental Protection Agency
Washington, DC 20460

EPA Notification of Hazardous Waste Activity

Please refer to the *Instructions for Filing Notification* before completing this form. The information requested here is required by law *(Section 3010 of the Resource Conservation and Recovery Act).*

For Official Use Only

Comments

C
C

Installation's EPA ID Number | Approved | Date Received *(yr. mo. day)*

C T/A C
F 1

I. Name of Installation

II. Installation Mailing Address

Street or P.O. Box

C
3

City or Town | State | ZIP Code

C
4

III. Location of Installation

Street or Route Number

C
5

City or Town | State | ZIP Code

C
6

IV. Installation Contact

Name and Title *(last, first, and job title)* | Phone Number *(area code and number)*

C
2

V. Ownership

A. Name of Installation's Legal Owner | B. Type of Ownership *(enter code)*

C
R

VI. Type of Regulated Waste Activity *(Mark 'X' in the appropriate boxes. Refer to instructions.)*

A. Hazardous Waste Activity

☐ 1a. Generator ☐ 1b. Less than 1,000 kg/mo.
☐ 2. Transporter
☐ 3. Treater/Storer/Disposer
☐ 4. Underground Injection
☐ 5. Market or Burn Hazardous Waste Fuel *(enter 'X' and mark appropriate boxes below)*
☐ a. Generator Marketing to Burner
☐ b. Other Marketer
☐ c. Burner

B. Used Oil Fuel Activities

☐ 6. Off-Specification Used Oil Fuel *(enter 'X' and mark appropriate boxes below)*
☐ a. Generator Marketing to Burner
☐ b. Other Marketer
☐ c. Burner
☐ 7. Specification Used Oil Fuel Marketer *(or On site Burner)* Who First Claims the Oil Meets the Specification

VII. Waste Fuel Burning: Type of Combustion Device *(enter 'X' in all appropriate boxes to indicate type of combustion device(s) in which hazardous waste fuel or off-specification used oil fuel is burned. See instructions for definitions of combustion devices.)*

☐ A. Utility Boiler ☐ B. Industrial Boiler ☐ C. Industrial Furnace

VIII. Mode of Transportation *(transporters only — enter 'X' in the appropriate box(es)*

☐ A. Air ☐ B. Rail ☐ C. Highway ☐ D. Water ☐ E. Other *(specify)*

IX. First or Subsequent Notification

Mark 'X' in the appropriate box to indicate whether this is your installation's first notification of hazardous waste activity or a subsequent notification. If this is not your first notification, enter your installation's EPA ID Number in the space provided below.

☐ A. First Notification ☐ B. Subsequent Notification *(complete item C)*

C. Installation's EPA ID Number

EPA Form 8700-12 (Rev. 11-85) Previous edition is obsolete.

Continue on reverse

Figure 8.4

ID — For Official Use Only														
C													T/A	C
W														1

X. Description of Hazardous Wastes *(continued from front)*

A. Hazardous Wastes from Nonspecific Sources. Enter the four-digit number from 40 *CFR* Part 261.31 for each listed hazardous waste from nonspecific sources your installation handles. Use additional sheets if necessary.

1	2	3	4	5	6
7	8	9	10	11	12

B. Hazardous Wastes from Specific Sources. Enter the four-digit number from 40 *CFR* Part 261.32 for each listed hazardous waste from specific sources your installation handles. Use additional sheets if necessary.

13	14	15	16	17	18
19	20	21	22	23	24
25	26	27	28	29	30

C. Commercial Chemical Product Hazardous Wastes. Enter the four-digit number from 40 *CFR* Part 261.33 for each chemical substance your installation handles which may be a hazardous waste. Use additional sheets if necessary.

31	32	33	34	35	36
37	38	39	40	41	42
43	44	45	46	47	48

D. Listed Infectious Wastes. Enter the four-digit number from 40 *CFR* Part 261.34 for each hazardous waste from hospitals, veterinary hospitals, or medical and research laboratories your installation handles. Use additional sheets if necessary.

49	50	51	52	53	54

E. Characteristics of Nonlisted Hazardous Wastes. Mark 'X' in the boxes corresponding to the characteristics of nonlisted hazardous wastes your installation handles. *(See 40* CFR *Parts 261.21 — 261.24)*

☐ 1. Ignitable *(D001)*   ☐ 2. Corrosive *(D002)*   ☐ 3. Reactive *(D003)*   ☐ 4. Toxic *(D000)*

XI. Certification

I certify under penalty of law that I have personally examined and am familiar with the information submitted in this and all attached documents, and that based on my inquiry of those individuals immediately responsible for obtaining the information, I believe that the submitted information is true, accurate, and complete. I am aware that there are significant penalties for submitting false information, including the possibility of fine and imprisonment.

Signature	Name and Official Title *(type or print)*	Date Signed

EPA Form 8700-12 (Rev. 11-85) Reverse

Figure 8.4 (*cont.*)

Please print or type. *(Form designed for use on elite (12-pitch) typewriter.)*	*Form Approved. OMB No. 2050-0039. Expires 9-30-88*

UNIFORM HAZARDOUS WASTE MANIFEST

1. Generator's US EPA ID No. | Manifest Document No. | 2. Page 1 of | Information in the shaded areas is not required by Federal law.

GENERATOR

3. Generator's Name and Mailing Address
A. State Manifest Document Number
B. State Generator's ID
4. Generator's Phone ( )
5. Transporter 1 Company Name | 6. US EPA ID Number | C. State Transporter's ID | D. Transporter's Phone
7. Transporter 2 Company Name | 8. US EPA ID Number | E. State Transporter's ID | F. Transporter's Phone
9. Designated Facility Name and Site Address | 10. US EPA ID Number | G. State Facility's ID | H. Facility's Phone

11. US DOT Description *(Including Proper Shipping Name, Hazard Class, and ID Number)*	12. Containers No.	12. Containers Type	13. Total Quantity	14. Unit Wt/Vol	I. Waste No.
a.					
b.					
c.					
d.					

J. Additional Descriptions for Materials Listed Above
K. Handling Codes for Wastes Listed Above

15. Special Handling Instructions and Additional Information

16. **GENERATOR'S CERTIFICATION:** I hereby declare that the contents of this consignment are fully and accurately described above by proper shipping name and are classified, packed, marked, and labeled, and are in all respects in proper condition for transport by highway according to applicable international and national government regulations.

If I am a large quantity generator, I certify that I have a program in place to reduce the volume and toxicity of waste generated to the degree I have determined to be economically practicable and that I have selected the practicable method of treatment, storage, or disposal currently available to me which minimizes the present and future threat to human health and the environment; **OR,** if I am a small quantity generator, I have made a good faith effort to minimize my waste generation and select the best waste management method that is available to me and that I can afford.

Printed/Typed Name | Signature | *Month Day Year*

TRANSPORTER

17. Transporter 1 Acknowledgement of Receipt of Materials
Printed/Typed Name | Signature | *Month Day Year*

18. Transporter 2 Acknowledgement of Receipt of Materials
Printed/Typed Name | Signature | *Month Day Year*

FACILITY

19. Discrepancy Indication Space

20. Facility Owner or Operator: Certification of receipt of hazardous materials covered by this manifest except as noted in Item 19.
Printed/Typed Name | Signature | *Month Day Year*

EPA Form 8700-22 (Rev. 9-86) Previous editions are obsolete.

Figure 8.5

Please print or type. *(Form designed for use on elite (12-pitch) typewriter.)* *Form Approved. OMB No. 2050-0039. Expires 9-30-88*

UNIFORM HAZARDOUS WASTE MANIFEST

1. Generator's US EPA ID No. Manifest Document No.

2. Page 1 of

Information in the shaded areas is not required by Federal law.

3. Generator's Name and Mailing Address

A. State Manifest Document Number

B. State Generator's ID

4. Generator's Phone ( )

5. Transporter 1 Company Name

6. US EPA ID Number

C. State Transporter's ID

D. Transporter's Phone

7. Transporter 2 Company Name

8. US EPA ID Number

E. State Transporter's ID

F. Transporter's Phone

9. Designated Facility Name and Site Address

10. US EPA ID Number

G. State Facility's ID

H. Facility's Phone

GENERATOR

11. US DOT Description *(Including Proper Shipping Name, Hazard Class, and ID Number)*	12. Containers No.	12. Containers Type	13. Total Quantity	14. Unit Wt/Vol	I. Waste No.
a.					
b.					
c.					
d.					

J. Additional Descriptions for Materials Listed Above

K. Handling Codes for Wastes Listed Above

15. Special Handling Instructions and Additional Information

16. GENERATOR'S CERTIFICATION: I hereby declare that the contents of this consignment are fully and accurately described above by proper shipping name and are classified, packed, marked, and labeled, and are in all respects in proper condition for transport by highway according to applicable international and national government regulations.

If I am a large quantity generator, I certify that I have a program in place to reduce the volume and toxicity of waste generated to the degree I have determined to be economically practicable and that I have selected the practicable method of treatment, storage, or disposal currently available to me which minimizes the present and future threat to human health and the environment; OR, if I am a small quantity generator, I have made a good faith effort to minimize my waste generation and select the best waste management method that is available to me and that I can afford.

Printed/Typed Name | Signature | *Month Day Year*

TRANSPORTER

17. Transporter 1 Acknowledgement of Receipt of Materials

Printed/Typed Name | Signature | *Month Day Year*

18. Transporter 2 Acknowledgement of Receipt of Materials

Printed/Typed Name | Signature | *Month Day Year*

FACILITY

19. Discrepancy Indication Space

20. Facility Owner or Operator: Certification of receipt of hazardous materials covered by this manifest except as noted in Item 19.

Printed/Typed Name | Signature | *Month Day Year*

EPA Form 8700-22 (Rev. 9-86) Previous editions are obsolete.

Figure 8.5 (*cont.*)

If more than 100 kilograms (220 pounds), but less than 1,000 kilograms (2,200 pounds), of waste are generated in a single month, the manifest requirements may not apply. They do not apply under the following conditions.

- If the waste is reclaimed under a contractual agreement that identifies the type of waste and the frequency of shipments
- If the reclaimer of the waste owns the vehicle that transports the waste
- If the generator maintains a copy of the reclamation agreement for at least three years after the agreement terminates

Once the EPA or the state Manifest form has been completed by the generator, he or she must obtain the handwritten signature of the initial transporter and the date of acceptance of the Manifest. The generator must reproduce enough copies of the Manifest to provide the generator, each transporter, and the owner or operator of the designated facility with one copy each for their records, and another copy to be returned to the generator. When shipment is made, the generator retains the top copy of the Manifest and gives the initial transporter the remaining copies. The regulations also require a specified number of copies of the Manifest which must be distributed for water or rail shipments of hazardous waste within the United States.

Pre-Transport Requirements

Before a generator may transport hazardous waste or contract for transportation of such waste off-site, the waste must be packaged, labeled, marked, and provided with placards in accordance with Department of Transportation regulations for hazardous waste. In addition, containers holding 110 gallons or less of hazardous waste must have the following words and information clearly displayed on the container.

> HAZARDOUS WASTE – *Federal law prohibits improper disposal. If found, contact the nearest police or public safety authority or the* U.S. *Environmental* Protection *Agency.*
> *Generator's Name and Address*
> *Manifest Document Number*

Figure 8.6 is a copy of a commonly used, commercially available, self-sticking placard that meets this requirement.

The pre-transport stipulations also state specific time limits during which a generator may accumulate hazardous waste on its premises. These requirements permit generators to accumulate hazardous waste prior to transport without being viewed as participating in *storage*, and thus without being required to obtain a storage permit from the EPA. As a general rule, generators may accumulate a hazardous waste for 90 days or less without a permit, so long as the requirements shown in Figure 8.7 are met.

If a generator accumulates hazardous waste for more than 90 days, he or she is deemed to be an operator of a storage facility and is subject to those requirements applicable to owners or operators of storage facilities. However, the regulations provide that if the waste has remained on site for longer than 90 days due to unforeseen, temporary, and uncontrollable circumstances, then the EPA may, at its discretion, grant a 30-day extension.

Generators are permitted to accumulate up to 55 gallons of hazardous waste, or one quart of acutely hazardous waste, without an accumulation time limit or a permit, provided the waste is kept in properly designed and maintained containers located at or near the point of generation. The operator of the process generating the waste must be in control of the container holding the waste, and the generator must clearly mark the containers with the words "Hazardous Waste." If the generator accumulates more than 55 gallons of hazardous waste or one quart of acutely hazardous waste, then the generator must, within three days after accumulating the excess amount, comply with the general requirements for accumulating hazardous wastes for 90 days or less.

Generators who produce more than 100 kilograms (220 pounds), but less than 1,000 kilograms (2,200 pounds) of hazardous waste in a calendar month are permitted to accumulate such waste for up to 180 days without a permit if they comply with a number of specific requirements. First, the total accumulated quantity of waste must never exceed 6,000 kilograms (13,200 pounds). Next, the generator must comply with most of the *Use and*

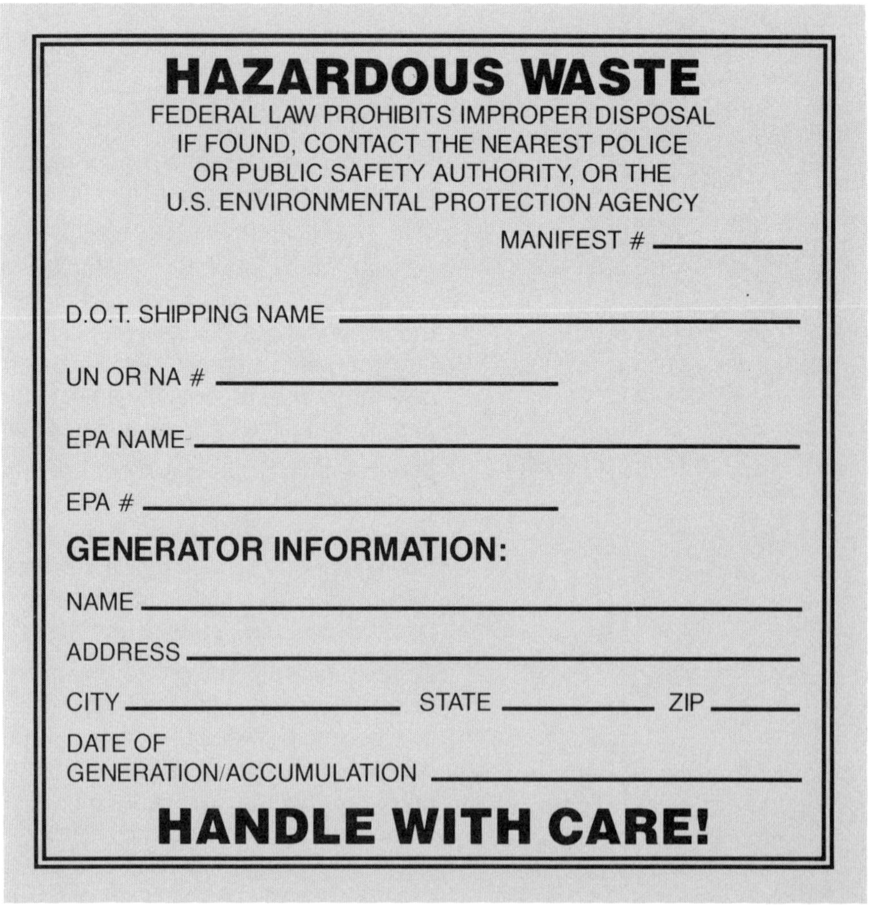

HAZARDOUS WASTE

FEDERAL LAW PROHIBITS IMPROPER DISPOSAL
IF FOUND, CONTACT THE NEAREST POLICE
OR PUBLIC SAFETY AUTHORITY, OR THE
U.S. ENVIRONMENTAL PROTECTION AGENCY

MANIFEST # ________

D.O.T. SHIPPING NAME ________

UN OR NA # ________

EPA NAME ________

EPA # ________

GENERATOR INFORMATION:

NAME ________

ADDRESS ________

CITY ________ STATE ________ ZIP ________

DATE OF
GENERATION/ACCUMULATION ________

HANDLE WITH CARE!

Figure 8.6

Management of Containers regulations and the *Tank Systems* regulations applicable to 100-to-1,000 kilograms per month hazardous waste generators. Furthermore, the containers for such waste must be properly labeled in accordance with the pre-transport requirements. In addition, the generator must comply with a number of emergency response measures.

If a generator producing between 100 and 1,000 kilograms of hazardous waste per month must have its waste transported more than 200 miles for off-site treatment, storage, or disposal, he or she is permitted to maintain the hazardous waste on-site for 270 days, rather than 180 days.

Finally, any generator of hazardous waste (regardless of whether he or she qualifies for a 180-day or 270-day accumulation period) who accumulates more than 6,000 kilograms of hazardous waste is considered an *operator of a storage facility* unless he or she applies for, and is granted, a 30-day extension to the accumulation period for unforeseen, temporary, and uncontrollable circumstances.

Record Keeping and Reporting

Generators are required to maintain signed copies of each Manifest, all test and analysis records, and all *Biannual Reports* and *Exception Reports* for a period of at least three years. The *Biannual Report* is a report submitted on EPA Form 8700-13A. It is prepared by the generator and identifies his or her activity during a previous year. It must contain the EPA identification numbers for the generator; the transporters and the off-site treatment, storage, or disposal facilities; the EPA hazardous waste number(s); and a description of the volume and toxicity of the waste generated. A copy of EPA Form 8700-13A is in Appendix E.

If within 35 days after the waste is delivered by a generator to an initial transporter, the generator has not received a return copy of the Manifest (with the handwritten signature of the owner or operator of the designated disposal facility), the generator is obligated to contact the transporter or the owner or operator of the designated disposal facility to determine the status of the waste. If within 45 days after shipment, the generator has still not

Requirements for Short-term Storage of Hazardous Waste

1. Waste must be stored in appropriate containers.
2. The containers must be clearly marked with the words "Hazardous Waste."
3. The date upon which accumulation began must be shown on the container.
4. The container must be properly maintained.
5. The generator must have a "Preparedness Plan," a "Contingency Plan," and an "Emergency Procedures Plan" on-site to deal with spills.
6. Site personnel must be properly trained to implement the contingency and emergency plans.

Figure 8.7

received this copy of the Manifest, he must submit what is known as an *Exception Report* to the EPA Regional Administrator for the region in which the generator is located. The Exception Report must include a copy of the manifest and a cover letter from the generator explaining all efforts undertaken to locate the hazardous waste.

Note that the reporting requirements exempt generators of 100 to 1,000 kilograms per month from all reporting requirements other than for the Manifest and for maintaining records of tests and analyses conducted.

Standards Applicable to Transporters

Transporters of hazardous wastes are required to obtain EPA identification numbers, comply with the manifest requirements, and maintain detailed records. Furthermore, the regulations dictate the actions that must be taken when a discharge (i.e., spill) of hazardous waste occurs. Such procedures include the cleanup of such discharges. The RCRA transportation regulations also incorporate a number of pertinent Department of Transportation regulations pertaining to transporters of hazardous waste.

Standards for Owners and Operators of Treatment, Storage, and Disposal Facilities

EPA has issued extensive regulations pertaining to the ownership and operation of hazardous waste treatment, storage, and disposal facilities. The purpose of these regulations is to establish minimum national standards to define the acceptable management of hazardous waste. These standards apply to a broad category of owners and operators.

General Requirements

A number of general requirements are set forth initially. These include:

- Mandatory application for EPA identification numbers
- Various notification requirements pertaining to the shipment of hazardous waste to the facility from off-site and foreign sources, and for the transfer of ownership or operation of a facility
- Waste analysis requirements
- Security maintenance requirements
- Inspection requirements
- Facility personnel training
- Facility location standards

Emergency Measures

A separate sub-part of the owner/operator regulations is dedicated to preparedness for, and prevention of, emergencies, such as fire, explosion, or release of hazardous waste. That section sets forth:

- Facility design and operation standards
- Equipment requirements
- Testing and maintenance requirements
- Requirements for access to communications and alarm systems
- Demands and efforts at coordination with local authorities

Contingency Plans

Sub-part "D" of the RCRA regulations is dedicated to the establishment of contingency plans, and emergency procedures for owners and operators of storage, treatment, or disposal facilities. The regulations require that all owners and operators of hazardous waste facilities maintain a contingency plan and that the plan be comprised of specific elements. Owners and operators are also obligated to follow a number of emergency procedures which not only include procedures for actually attending to the emergency, but also set forth extensive notification requirements. Figure 8.8 lists those requirements applicable to generators of hazardous waste, such as contractors.

Contingency Plan Requirements:

Preparing for and Preventing Accidents

- Install and maintain emergency equipment such as alarms, telephones, or two-way portable radios, fire extinguishers (using water, foam, inert gas, or dry chemicals as appropriate to your waste type), hoses, automatic sprinklers, or spray equipment immediately available to employees for emergency use.
- Provide enough room for emergency equipment and response teams to get to any area of the project in the event of an emergency.
- Contact local fire, police, and hospital officials or state or local emergency response teams and explain the types of wastes expected to be generated. Ask for their cooperation and assistance in handling emergency situations.
- Maintain emergency telephone number lists next to emergency communication equipment.
- Appoint an employee to act as emergency coordinator to ensure that emergency procedures are carried out in the event an emergency arises. The responsibilities of the emergency coordinator are generally that he/she be available 24 hours a day (at the facility or by phone) and know whom to contact and what steps to follow in an emergency.

Planning for Emergencies

In the event of an emergency:

1. In the event of a fire, call the fire department or attempt to extinguish it using the appropriate type of fire extinguisher.
2. In the event of a spill, contain the flow of hazardous waste to the extent possible and notify the National Response Center. The Center operates a 24-hour toll-free number: 800-424-8802, or in Washington, D.C.: 426-2675. As soon as possible, clean up the hazardous waste and any contaminated materials or soil.
3. In the event of a fire, explosion, or other release, immediately notify the National Response Center as required by Superfund regulations. (Superfund is the law that deals with the cleanup of spills and leaks of hazardous waste at abandoned hazardous waste sites.)

Figure 8.8

Manifest, Record Keeping, and Reporting Requirements

As with generators and transporters, owners and operators of hazardous waste treatment and disposal facilities are subject to extensive Manifest, record keeping, and reporting requirements. In addition, there are numerous regulations pertaining to the prevention and detection of releases from surface impoundments, waste piles, and other land treatment areas or landfills. These regulations cover:

- The establishment of concentration limits for hazardous constituents
- General groundwater monitoring requirements
- Minimum requirements for detection monitoring programs
- Compliance with monitoring programs and corrective action programs

In accordance with RCRA, the EPA has issued voluminous regulations pertaining to the operation of various types of hazardous waste treatment, storage, and disposal facilities. Entire sub-parts have been set aside to address closure and post-closure requirements, financial requirements, the use and management of hazardous waste containers, the use of tank systems, surface impoundments, waste piles, land treatment, landfills, and incinerators. These detailed and technical requirements span almost 100 pages of regulations.

Authorized State Hazardous Waste Programs

Congress has granted the EPA authority under RCRA to issue guidelines to assist states in the development of hazardous waste programs. RCRA also grants EPA the authority to approve the implementation of state hazardous waste programs. Such authorization entitles the state to administer and enforce its own program in lieu of the federal program and to issue and enforce permits for the storage, treatment, or disposal of hazardous waste. In addition, RCRA provides standards, requirements, and procedures for EPA authorization of state programs, and provides a mechanism by which EPA can "police" state implementation of hazardous waste programs.

Miscellaneous Provisions

RCRA grants the EPA and state authorities broad powers over hazardous waste inspections. Only designated EPA officers, employees, and representatives, and designated state officers, employees, and representatives (in states that have authorized hazardous waste programs) have the authority to carry out inspections. Premises may be inspected where hazardous wastes have been generated, stored, treated, disposed, or shipped. Samples of such wastes may be obtained and inspected. Records, reports, permits, or other information pertaining to such wastes may also be inspected.

Federal Enforcement

Importantly, the EPA has been granted strict federal enforcement authority under RCRA. Whenever the EPA determines that any person has violated or is in violation of any requirement of this subchapter, the Administrator may issue an order assessing a civil penalty for any past or current violation, and may require compliance immediately, or within a specified time period. Alternatively, the Administrator may commence a civil action in the United States District Court in the district in which the violation occurred, seeking appropriate relief, including a temporary or permanent injunction.

Civil penalties may include suspension or revocation of permits issued by the EPA or by states. Where money damages are appropriate, such penalties may not exceed $25,000 per day of noncompliance.

Before an order is issued or a civil action commenced under RCRA, the EPA must notify the state in which the violation occurred if that state has been authorized to carry out its own hazardous waste program. Persons to whom violation orders are issued have 30 days in which to request a public hearing. If such a request is not made within the 30-day period, the orders are deemed final. By the same token, requests for a public hearing must be acted upon promptly by EPA. Pre-hearing and hearing procedures are similar to those employed in preparation for and in the conduct of other administrative hearings or trials.

Finally, RCRA sets forth maximum penalty amounts and criminal penalties, including imprisonment terms, for knowingly violating almost any RCRA requirement. Upon criminal conviction for a knowing violation that does not place another person in imminent danger of death or serious bodily injury, an individual will be subject to a fine of not more than $50,000 for each day of the violation and/or imprisonment not to exceed either two years or five years (depending on the violation). Where the violation subjects another person to imminent danger of death or serious bodily injury, the responsible party may be subject to a fine of not more than $250,000 and/or imprisonment for not more than 15 years. For violations not enumerated in RCRA, violators may be subject to the imposition of a civil penalty in an amount not to exceed $25,000 for each day of the violation.

Monitoring, Analysis, and Testing

Under RCRA, the EPA has the authority to issue an order requiring an owner or operator of a facility or site to conduct monitoring, testing, analysis, and reporting with respect to that facility or site if the EPA determines that the presence of any hazardous waste or the release of any such waste may pose a substantial hazard to human health or the environment. In addition, the EPA may also order such monitoring, testing, analysis, and reporting by an owner or an operator of a facility even if the site is no longer in operation at the time the EPA determines that such testing is necessary.

If the EPA determines that there is no owner or operator capable of conducting such testing, or if it determines that such testing has been conducted in an unsatisfactory manner, the EPA may conduct the testing itself or authorize a state or local authority or other persons to do so. It may then seek reimbursement from the owner or operator for all costs incurred.

Finally, if any person fails to comply with any order to undertake monitoring, analysis and testing, the EPA may commence a civil action against such individual, seeking compliance with its order and/or the imposition of a civil penalty not to exceed $5,000 for each day during which the failure continues.

Notification Requirements

RCRA sets forth a notification requirement applicable to all generators, transporters, and facility owners or operators. Whenever a generator or transporter, owner, or operator commences activities in which he or she is handling hazardous waste, the EPA, or an authorized state representative, must be notified, using the EPA's standard notification form for providing such notice. A generator or transporter, owner, or operator who fails to provide this notice will be barred from continuing their hazardous waste-related activities. Figure 8.4 is a copy of the standard EPA Notification Form.

Miscellaneous Provisions

Employee Protection

Section 6971 of RCRA, entitled "Employee Protection," bars any person from firing or discriminating against an employee (or authorized representative of an employee) because that employee (or representative) has instituted proceedings (under RCRA or any applicable implementation plan), or has testified or plans to testify in any proceedings (concerning the enforcement of any of RCRA's provisions or of any applicable implementation plan). This section also sets forth procedures by which an employee or representative who suspects that he or she has been fired or discriminated against may seek review of the alleged wrongful action by the Secretary of Labor. The Secretary of Labor's investigation of such alleged wrongful acts must include the opportunity for a public hearing. Where an order is issued by the Secretary of Labor finding discrimination or wrongful firing, the employee or representative is entitled to reimbursement for all costs and expenses, including attorney's fees. Any order issued by the Secretary of Labor under this section may be appealed for judicial review.

Citizen Suits

Any citizen is entitled to file a lawsuit against any other person, the United States, the EPA, and other government agencies for violations of any permit, standard, regulation, condition, requirement, prohibition, or order issued pursuant to RCRA. Citizens wishing to file such suits, known as *citizen suits*, must, as a general rule, provide 60 days prior notice of the violation to the EPA, the state in which the alleged violation occurred, and to the alleged violator. Where there has not been a specific statutory or regulatory violation, but where any present or past generators, transporters, owners or operators have contributed to, or are contributing to, the handling, storage, treatment, transportation or disposal of any solid or hazardous waste (which may present an imminent and substantial danger to human health or the environment), a citizen is entitled to file a suit against such generators, transporters, owners, or operators after providing 90 days prior notice to the EPA of the state in which the alleged endangerment occurred, and the alleged violator. Where a citizen alleges a violation of the provisions of Subchapter III, the Hazardous Waste Management provisions of RCRA, no prior notice is required. All citizen suits must be filed in the United States District Court and if filed against the EPA, the action must be filed in the District Court for the District of Columbia.

The citizen suits provisions of RCRA do prohibit lawsuits in a few instances. For example, actions pertaining to alleged imminent and substantial endangerment may not be filed if they involve activities to which the EPA or the state are contributing. Actions that involve the siting of a hazardous treatment, storage, or disposal facility are also prohibited. Furthermore, a citizen may not seek to enjoin the issuance of a facility permit.

Imminent Hazard

RCRA contains a special provision that is common to a number of environmental statues. It provides EPA with the authority to file a lawsuit on behalf of the United States against past and present generators, transporters, owners, or operators where there is evidence that "the past or present handling, storage, treatment, transportation or disposal of any solid waste or hazardous waste may present an imminent or a substantial endangerment to health or to the environment." EPA's authority under RCRA is similar to its authority under the Comprehensive Environmental Response, Compensation, and Liability Act (more commonly known as "Superfund").

The Imminent Hazard Provision is a powerful tool with which the EPA can impose cleanup liability on all persons who have at any time contributed to the handling, treatment, storage, transportation, or disposal of waste, and such actions have led to an imminent and substantial endangerment. Such persons include the general contractor whose wastes are not properly disposed, even if that general contractor hired someone else to take care of that problem for him. Any person found to have caused or contributed to an imminent hazard is subject to either an order restraining him or her from the hazardous activity, *or* an order directing him or her to take any other action necessary to effect remediation of the hazard. Violators of either order are subject to a fine of $5,000 per day of continued violation.

Regulation of Underground Storage Tanks

As part of the 1984 RCRA amendments, Congress issued a completely new subchapter to the statute. Subchapter IX, entitled "Regulation of Underground Storage Tanks," grants the EPA, for the first time, the authority to regulate the management of underground storage tanks. This new subchapter was the result of increased recognition that leaks from unsupervised and unregulated underground storage tanks have become serious environmental threats.

This new section of RCRA, and the regulations issued thereunder, apply to any one or any combination of tanks (including underground pipes connected thereto) used to contain an accumulation of regulated substances, and the volume of which (including the volume of the underground pipes connected thereto) is ten percent or more beneath the surface of the ground. The law exempts from its definition of *underground storage tank* a number of specific types of storage tanks (i.e., septic tanks, surface impoundments, pits, ponds, lagoons, and stormwater or wastewater collection systems).

Subchapter I lists numerous requirements for the maintenance and operation of underground storage tanks with which owners and operators of underground storage tanks must comply. *Owners* are defined as persons who own underground storage tanks used for the storage, use, or dispensing of regulated substances. *Operators* are defined as persons in control of or having responsibility for the daily operation of the underground storage tank. Owners are obligated to notify the designated state or local agency of the existence of underground storage tanks, and to provide a description of the tanks and of the wastes stored therein.

Moreover, the EPA is required to promulgate regulations for release, detection, prevention, and storage. These regulations apply to all owners and operators of underground storage tanks. The underground storage tank provisions also establish financial responsibility of the owners and operators, and specify new tank performance standards. Congress has provided specific performance standards in the text of the statute to serve as interim codes until the performance standards regulations have been promulgated.

Finally, Subchapter IX grants EPA the ability to authorize state underground storage tank release, detection, prevention, and correction programs. It further establishes a federal enforcement program for violations of any requirements of Subchapter IX.

Chapter 9

Hazardous Waste Management Planning

Preplanning for the safe management of hazardous wastes protects employees from health hazards, protects employers from exposure to liability, and protects the environment from contamination. This chapter provides guidance for developing an effective hazardous waste management plan.

The first section of the chapter deals with the categories of hazardous waste generators and what those categories mean. It also defines *hazardous wastes* in the context of construction sites. The second section outlines the development and contents of an effective plan for hazardous waste management. The various elements of the plan are discussed in greater detail in later chapters.

Categories of Hazardous Waste Generators

There are three categories used to classify hazardous waste generators: *Large, Small* and *Very Small*.

Large Quantity Generator (LQG): produces more than 1,000 kilograms (2,200 lbs.) of hazardous waste per month, or more than 1 kilogram (2.2 lbs.) of acutely hazardous waste per month.

Small Quantity Generator (SQG): produces more than 100 kilograms (220 lbs.) of hazardous waste per month, but less than 100 kilograms, and less than 1 kilogram (2.2 lbs.) of acutely hazardous waste per month.

Very Small Quantity Generator (VSQG): produces less than 100 kilograms (220 lbs.) of hazardous waste per month and no acutely hazardous waste.

Hazardous Waste Definition

The term *hazardous waste* is interpreted in different ways by a variety of statutes. No one statute is comprehensive in its definition. Many of the statutes list specific chemicals or substances which are hazardous by definition and which, if disposed, are hazardous wastes. Other statutes try to define hazardous wastes by category or characteristic.

The term is used in this book to mean any substance that is (1) no longer needed, and (2) defined as hazardous by statute, or likely to cause an adverse health effect in humans.

Examples of Hazardous Wastes

Figure 9.1 identifies a number of hazardous materials commonly found on construction sites. The materials shown are classified according to the characteristic which causes them to be hazardous.

Hazardous Waste Management Plan

The Planning Steps

Effective management of hazardous wastes requires careful pre-planning. The completeness of the plan will determine how effective it is. The following steps should be used to develop an effective plan. Each will be fully explained in subsequent paragraphs.

1. Notify EPA and state agencies.
2. Conduct a hazardous waste audit.
3. Accumulate MSDS forms and manifest data.
4. Evaluate waste handling, storage and disposal procedures.
5. Develop safe waste handling, storage, and disposal procedures.
6. Conduct training programs for personnel.
7. Implement the management plan.

Notification of EPA and State Agencies

The use of hazardous materials in the workplace almost always results in the generation of hazardous wastes. Current federal regulations require that EPA be notified whenever hazardous wastes are generated. Notification is not dependent upon the quantity of wastes generated, only the *fact* of generation. Thus, even Very Small Quantity generators are required to notify EPA of their waste generation activities.

In addition, most states require that the state environmental agency also be notified of hazardous waste generation within their borders. Both the state and federal notification are done at the same time using a single federal form.

Hazardous waste generation activities are site-specific. A contractor working on three identical projects at three different sites is required to file three separate notifications.

Hazardous waste registration forms require a listing of the wastes generated. The identification is by generic code established by EPA. Codes, which identify the general nature of the wastes, are established for nearly every waste conceivable and include catch-all categories for those wastes not otherwise defined. The codes are found in the codebook for EPA *Form* 8700-13A. Those in effect as of this writing are found in Appendix A.

To identify codes on the notification form requires the user to consider what wastes will be generated by the proposed activity. Therefore, as a particular job is being planned, it is useful to consider what materials will be used during the various phases, whether those materials are hazardous, and whether a waste product, including the "empty" containers, will be generated. That information alone is often adequate to file the appropriate notification forms.

In addition to state and federal notification requirements, many localities also require notification of the local Board of Health, Police Department, Fire Department, or Hazardous Waste Coordinator whenever hazardous waste generation activities are undertaken in the community. Contractors should make it a practice to contact the local Board of Health regarding notification requirements as part of their normal building permit application activities.

Conducting a Facility Audit

An audit is not intended to solve, fix, or cure anything. It does, however, identify those areas that may need attention. It points out improper or unsafe activities, and provides the foundation for developing effective hazardous

Hazardous Material Found on a Construction Site (partial list)		
Category	**Substance**	**Use Or Location**
Acids	Muriatic acid (Hydrochloric Acid)	Masonry cleaning
	Battery acid	Motor vehicle and equipment batteries
Alkalis	Portland Cement	Mortar, concrete
	Calcium hydroxide	Mortar additive
	Sodium hydroxide	Cleanser additive
	Calcium oxide	Drilling fluids, lubricants
	Lime	Landscaping activities
	Calcium chloride	Dust suppressant
	Lye	Detergent, cleaners
Asbestos	Shingles	Hard exterior shingles, roofing shingles
	Pipe insulation	Steam heating systems
	Fire retardants	Sprayed on structural members
	Drapes, carpets, ceiling tiles	Throughout buildings
Chlorinated Solvents	Degreasers	Motor pool areas
	Cleaners	Metalwork cleaning, polishing
	Paint removers (methylene chloride)	Rehabilitation work
Chlorine	Hypochlorite crystals	Water pipe disinfectant
	Liquid chlorine	Usually mixed on-site from crystals
	Gaseous chlorine	Water pipe disinfectant
Compressed Gases	Oxygen	Metal cutting, welding
	Acetylene	Metal cutting, welding
	Carbon dioxide	Pressure testing
	Air	Dust blowing
	Chlorine	Disinfection
	Hydrogen	Metal cutting, welding
Coolants	Antifreeze	Motor pool areas, trucks, vehicles, and heavy equipment radiators
	Freon	Refrigeration systems, air conditioning systems, chilled water systems
Dusts	Wood dust	Sawing, sanding wood
	Cement dust	Portland Cement or mortar batch mixers
	Wallboard dust	Taping, sanding, cutting, demolishing
	Metal dust	Cutting, grinding, sanding
	Paint dust	Sanding painted surfaces
	Lime dust	Landscaping operations
	Calcium dust	Dust control operations
	Soil dust	Excavation, backfilling, grading

Figure 9.1

Hazardous Material Found on a Construction Site (partial list) (continued)		
Category	**Substance**	**Use Or Location**
Flammable Solvents	Degreasers	Motor pool areas, metal-cleaning sprays
	Paint thinners	All finishing operations
	Cleaners	Paint and glue removers
	Carburetor cleaners	Motor pool areas
Fuels	Gasoline	Motor vehicles, equipment
	Diesel fuel	Motor vehicles, generator sets
	Kerosene	Space heaters
	White gas	Space heaters
Paints	Gallon cans	Finishing operations
	Spray cans	Motor pools, touch-up work
PCB's	Oils, lubricants	Electrical transformers, electrical capaciters in pre-1979 electrical equipment
Phenolic Compounds	Sealants	Painting foundation walls
	Waterproofing	Painting or spraying on foundations
	Cements	Roofing tars
	Wood Preservatives	Deck and sill painting
	Creosote	Foundation waterproofing
	Coal Tar	Disinfection
	Asphalt	Paving operations
Oils and Lubricants	Lubrication greases	Motor pool areas
	Form release compounds	Concrete forms
	Oils	Motor pool areas

Figure 9.1 (*cont.*)

waste management planning. In fact, a project that has been well planned in advance will not require an audit, except as a check on implementation of the plan.

Regardless of the reason for the audit, it is conducted in the same way. The auditor begins at one end of the project and systematically walks through the site, carefully looking at every storage, handling, use, and disposal activity being conducted. Using a form similar to that shown in Figure 9.2, the auditor records the nature of the waste generated by that activity, the intermediate disposal practice, and the ultimate means of disposal. When an activity is identified for which this information is not readily apparent, it is up to the auditor to find the answer.

Along with the form in Figure 9.2, the auditor should carry a note pad to record such information as quantities of waste that seem unusual; the close storage of incompatible wastes; and opportunities to reduce waste generation, recycle materials, or segregate waste streams for easier disposal.

Accumulation of MSDS Forms and Manifest Data

Under current law, the contractor is responsible for the ultimate fate of all hazardous wastes generated on the construction site. This means that the contractor must properly manifest (file the appropriate EPA form) all hazardous wastes disposed from the site. Proper manifesting requires an understanding of the nature of the wastes being disposed so that safe disposal can be assured.

The best source of information regarding the nature of the waste is the MSDS form for the originating material. Hazardous materials generally yield hazardous wastes. The waste manifest, however, only requires an identification of the hazard from a broad waste-compatibility standpoint. As expected, the manifest form uses codes to classify the wastes.

Manifest codes are generic. They depend upon broad ranges of reactivity, flammability, and toxicity. Classifying the wastes from multiple material classification codes into fewer waste classification codes is accomplished by applying the most conservative (i.e., most hazardous) information from the MSDS forms to the waste classification criteria. The criteria and codes are found in EPA Form 8700-13A. Those codes in use on the date of publication of this book are found in Appendix A and Appendix B. The codes in Appendix A classify the wastes, while the codes in Appendix B classify the generator.

Developing Appropriate Waste Management Practices

The manner in which hazardous wastes are handled, stored, and disposed is dictated by the nature of the waste. Chapters 10-12 describe specific practices to be used for various substances. The management plan should incorporate all those practices associated with the various wastes generated on the site.

To facilitate plan development, it is often useful to classify the various waste streams by compatibility code, and to plot points of generation on a facility site plan. Compatible wastes generated in close proximity should be stored together. The fewer waste storage locations, the safer the job site will be.

When selecting storage sites, the manner in which the wastes are transported should be considered. For example, if long distances must be covered, and the route goes by regular refuse disposal areas, there will be a strong temptation to improperly dispose hazardous wastes.

Subcontractor activities that will generate hazardous wastes also need to be considered during plan development. It is the responsibility of the general contractor to ensure safe waste management by its subcontractor. Making waste management practices easy to implement encourages compliance. A management plan prepared in advance of a project should include a policy

Site Inventory Form

1. WORK AREA	2. ROOM NO.	3. NATURE OF ACTIVITY	4. SURVEYOR	5. DATE	6. SHEET __ of __

7. MATERIAL	8. MANUFACTURER	9. MFGR. ADDRESS	10. LABEL?	11. M or W?	12. USE	13. FATE	14. WASTE FATE	15. MSDS?

Figure 9.2

statement that encourages plan utilization through penalties and incentives. Subcontracts should include clauses requiring the subcontractor to provide a hazardous waste management plan consistent with the general contractor's plan. These measures will create an awareness of the need to practice safe waste management and will force the subcontractor to think through his own waste disposal plan in advance of a problem.

Staff Training

The best laid plan will fail if those who must carry it out do not know about or understand it. The contractor should prepare and implement a training program for his management staff. The staff must then train employees regarding their responsibilities under the plan. Subcontractor personnel should be included in the training sessions where appropriate.

The training program should consist of the purpose of the plan, the specific practices required, the location of hazardous waste storage facilities, spill prevention and control procedures, emergency telephone numbers, and the location on the job site of the complete written plan. The elements of a typical plan are further described below.

Plan Implementation

It is the general contractor's responsibility to ensure that the plan is implemented by all of the firms' employees and subcontractors. The site superintendent is generally asked to work with the subcontractor to ensure compliance.

Typical Management Plan Contents

A hazardous waste management plan incorporating the concepts just discussed will typically follow an outline similar to that below. Each plan is site-specific, however. Therefore, some of the categories shown below will not apply to all sites, and other categories not shown may apply to particular sites. This outline should be used only as a guide to help arrange the plan for each site.

I. Introduction
   A. Purpose
   B. Applicability
   C. Examples of Hazardous Waste
   D. Hazard Communication Program
II. Handling of Hazardous Wastes
   A. Use of Intermediate Containers
   B. On-Site Movement Procedures
   C. Use of Designated Storage Areas – Policy
III. Storage Areas
   A. Location
   B. Access
   C. Inventory Control
   D. Separation of Wastes
   E. Responsibilities
   F. Policy
IV. Disposal
   A. Responsibilities
   B. Policy
   C. Procedures
V. Spill Prevention
   A. Policy
   B. Procedures

VI. Spill Cleanup
   A. Policy
   B. Procedures
VII. Incident Reporting
   A. Policy
   B. Procedures
VIII. Training
   A. Policy
   B. Responsibilities
   C. Procedures
IX. Labeling
   A. Policy
   B. Label description
   C. Labeling instructions

Chapter 10

Handling Hazardous Wastes

The risk to employees on a construction site may be greater from hazardous materials than from hazardous wastes. Nevertheless, actual injury occurs more often when handling hazardous wastes. The reason for this paradox is that wastes tend to be handled in a more cavalier manner than raw materials.

Most employees have an innate understanding of the risks associated with raw materials and have developed handling methods over time that minimize those risks. Unfortunately, most employees do not know or understand the risks associated with hazardous wastes; consequently, no safe handling methods have routinely evolved for such substances.

To minimize employee risk and company liability exposure, every employer should develop clear waste management guidelines for employees to follow. The guidelines must then be enforced. This chapter will address the issue of hazardous waste handling and will help employers develop the appropriate guidelines for their project(s).

Personal Protective Equipment

The need to use personal protective equipment (PPE) does not change when a substance changes from raw material to waste. If it was hazardous as a raw material, it will probably still be hazardous and maybe more dangerous—as a waste. PPE is discussed at length in Chapter 5. It is strongly recommended that those portions of Chapter 5 pertaining to equipment usage be reviewed in the context of hazardous waste.

Generally, gloves and aprons should be worn when handling any hazardous waste. Goggles are recommended when splashing of liquids is likely and when hazardous dust may be generated. A dust mask or respirator may be required if dust or fumes are present. The MSDS form which came with the raw material is the best source of information regarding the type of personal protective equipment to be used. The following sections prescribe the appropriate PPE for various applications.

Liquid Transfer

Most hazardous wastes are not inherently hazardous by their mere presence. Radioactive wastes and infectious wastes are exceptions; but those types of waste are seldom found on a construction site. Hazardous wastes endanger the worker only when a route of exposure is created, and actual exposure occurs.

During the handling of hazardous wastes, the worker is more at risk than at any other time. When the waste is in a liquid form, leakage, seepage, spills, and splashing create direct routes of exposure that may contaminate the worker.

Personal Protective Equipment

Whenever liquid hazardous wastes are being handled—whether they are being transported from the point of generation to a transfer container, to a storage area, or off-site for disposal—personal protective equipment must be utilized. The risk to the worker is very high during these activities, and appropriate protection is essential.

Pouring Liquids

It is the nature of liquids that they splash when being poured. However, there are ways to minimize splashing. When dealing with hazardous wastes, splash minimization is essential.

The first rule of pouring is to pour *slowly*. A slow, steady stream of liquid splashes far less than a "hard," erratic stream. This is easily demonstrated by turning on a tap in the sink. When the stream is slow, little splashing occurs. But when the faucet is open wide, splashing occurs everywhere. The same principle applies when pouring liquids into pails, bottles, or drums. A slow pour is a safe pour.

The second rule of pouring is to allow room for air to replace the liquid in the container being emptied. When containers have narrow mouths, it is easy to tip the container so fast that the entire mouth becomes clogged with liquid. This presents no problem for a second or two, but then the vacuum created inside the container by the evacuation of the liquid exceeds the air pressure at the mouth of the container. At that point a bubble pushes its way through the container opening to fill the vaccuum. The bubble briefly interrupts the flow of liquid, then allows it to resume until a second vacuum is created and the bottle "burps" again. That erratic "burping" is a major source of spills and splashing.

To avoid the "burps," pouring should be done slowly so that the mouth of the container is never fully filled with liquid. So long as air can pass over the stream of liquid into the container, "burping" will be eliminated.

The third rule of pouring is to remember that liquid displaces air in the receiving container. This is not a problem when pouring into an open-top drum or pail where the liquid stream entering the container is very much smaller than the container opening. When a funnel is used, however, the tendency is to jam the funnel into the container opening and then pour the liquids through the funnel. When this occurs, no room is left around the funnel through which the air can escape. Reverse "burping" occurs as a result. The liquid filling the container exceeds the outside air pressure. At that point, the air inside forces its way up the funnel tube to burst out of the top—usually carrying a splash of liquid with it. In addition, the funnel neck may become blocked with liquid and air, allowing the funnel to rapidly fill and overflow.

To avoid these problems, the funnel should be set into the opening *loosely*, or wedged with a small stick of wood, so that an air escape passageway is left between the funnel neck and the edge of the opening.

Carrying Containers

Liquid wastes should not be carried in open-top pails or glass bottles. Pails are easily spilled and glass breaks. The most suitable containers are those made from metal or plastic with tightly-fitted screw caps.

Drippings

Every time a liquid hazardous waste is transferred from one container to another, both containers should be carefully examined for evidence of drips or seepage. All drips or seepage found must be carefully wiped up with a rag or paper towel. The rag or towel should then be treated as a solid hazardous waste and disposed accordingly.

Spills

No matter how carefully workers follow handling procedures, spills are inevitable. Dry absorbent should be kept close at hand when transferring liquids so that spills can be quickly absorbed. Areas in which transfers occur regularly, such as the storage area, should have a layer of dry absorbent on the floor under the storage containers.

If a spill occurs during transport of the wastes between the point of generation and the point of storage, rapid containment and cleanup is essential. The procedures outlined in Chapter 6 for cleanup of hazardous material spills also apply to hazardous waste spills. Keep in mind that any cleanup material contaminated with hazardous waste also becomes a hazardous waste.

Solids Transfer

Solid hazardous wastes on construction sites usually take the form of cleanup materials used to control liquid spills, dusts from metal grinding operations, contaminated soils, and contaminated clothing. Automobile and truck batteries, oil filters, brake shoes, etc., are also included in this category. Containers for solid wastes generally do not need to be as sturdy as for liquids, because solid spills are contained and cleaned up more easily.

Personal Protective Equipment

The most likely routes of exposure to solid waste are inhalation and ingestion. Dusts are the most common form of solid hazardous wastes encountered. Breathing dusts, or touching the mouth with contaminated hands, are the ways in which dusts may enter the body. Dust masks, gloves, aprons, and eye goggles are usually adequate to protect the worker. The appropriate MSDS form should always be checked, however, to ensure proper levels of protection.

Sweeping Solids

Dust is generated in huge quantities by careless sweeping. Areas in which hazardous dusts (or any dusts, for that matter) are being swept should be lightly sprayed with clean water before the sweeping occurs. If that is not possible or practical, vacuum cleaning should be employed in lieu of sweeping.

Dumping Solids

The act of dumping solids creates heavy air drafts which lift large quantities of dust and allow them to be blown about the work site. Care is needed when dumping solid wastes to minimize dust generation. In particular, transfer containers can be rapidly inverted into the storage drum and set on the bottom of the drum or pile of waste. The transfer container can then be lifted off the refuse slowly so that the refuse does not drop into the drum. In addition, storage areas for solid wastes should be protected from the wind. Transfer during high winds should be avoided.

Chapter 11

On-Site Storage of Hazardous Wastes

All construction projects generate hazardous wastes. How those wastes are managed determines the liability exposure of the contractor and the risk exposure of his/her employees. There is a safe way to store hazardous wastes and there is a legal time limit on how long they may be stored. This chapter will address the safe and lawful storage of hazardous wastes. When used in conjunction with a proper Right-to-Know training program and hazard communication manual, the practices outlined in this chapter will maintain employee safety.

Note that all hazardous waste generators and storage facilities are required by federal law to notify EPA of the fact that they will be generating and/or storing hazardous wastes. In addition, they must identify the nature of those wastes and the quantities expected to be generated or stored. Failure to notify EPA in advance is a criminal offense.

Furthermore, most states also require a separate notification of the same activities. Some state laws define the lower limits of generation quantities much more rigorously than the federal government does. Thus, a generator/storer who generates quantities below the EPA notification threshold level may still be required to file notifications with the local or state governments. Failure to properly notify a state agency of generator/storage activities is also a criminal offense in most cases.

Notification of EPA is done by filing the standard form shown in Figure 11.1. Each state requiring separate notification usually has its own standard form that must be used. Copies of state forms can usually be obtained from the state environmental protection agency.

Compatibility/ Segregation

As with hazardous materials, hazardous wastes are not always compatible. Acidic wastes react violently with alkaline wastes, for example. Cyanide wastes generate toxic fumes when mixed with acidic wastes. Waste solvents can emit all manner of toxic fumes in the presence of acidic or alkaline wastes, depending upon the contaminants in the solvent. Some wastes generate extreme heat when mixed together, leading to fires or explosions. Even wastes that do not react (form toxic by-products when mixed) often need to be segregated, simply because the disposal problems are significantly more complicated with mixed wastes.

The guidelines provided in Chapter 5 for the segregation of hazardous materials generally apply to hazardous wastes as well. The waste should be

Please print or type with ELITE type *(12 characters per inch)* in the unshaded areas only

Form Approved. OMB No. 2050-0028. Expires 9-30-88.
GSA No. 0246-EPA-OT

United States Environmental Protection Agency
Washington, DC 20460

EPA Notification of Hazardous Waste Activity

Please refer to the *Instructions for Filing Notification* before completing this form. The information requested here is required by law *(Section 3010 of the Resource Conservation and Recovery Act).*

For Official Use Only

Comments

C
C

Installation's EPA ID Number — C F — T/A C 1 — Approved — Date Received *(yr. mo. day)*

I. Name of Installation

II. Installation Mailing Address

Street or P.O. Box — C 3

City or Town — C 4 — State — ZIP Code

III. Location of Installation

Street or Route Number — C 5

City or Town — C 6 — State — ZIP Code

IV. Installation Contact

Name and Title *(last, first, and job title)* — C 2 — Phone Number *(area code and number)*

V. Ownership

A. Name of Installation's Legal Owner — C R — B. Type of Ownership *(enter code)*

VI. Type of Regulated Waste Activity *(Mark 'X' in the appropriate boxes. Refer to instructions.)*

A. Hazardous Waste Activity

- ☐ 1a. Generator ☐ 1b. Less than 1,000 kg/mo.
- ☐ 2. Transporter
- ☐ 3. Treater/Storer/Disposer
- ☐ 4. Underground Injection
- ☐ 5. Market or Burn Hazardous Waste Fuel *(enter 'X' and mark appropriate boxes below)*
  - ☐ a. Generator Marketing to Burner
  - ☐ b. Other Marketer
  - ☐ c. Burner

B. Used Oil Fuel Activities

- ☐ 6. Off-Specification Used Oil Fuel *(enter 'X' and mark appropriate boxes below)*
  - ☐ a. Generator Marketing to Burner
  - ☐ b. Other Marketer
  - ☐ c. Burner
- ☐ 7. Specification Used Oil Fuel Marketer *(or On site Burner)* Who First Claims the Oil Meets the Specification

VII. Waste Fuel Burning: Type of Combustion Device *(enter 'X' in all appropriate boxes to indicate type of combustion device(s) in which hazardous waste fuel or off-specification used oil fuel is burned. See instructions for definitions of combustion devices.)*

☐ A. Utility Boiler ☐ B. Industrial Boiler ☐ C. Industrial Furnace

VIII. Mode of Transportation *(transporters only — enter 'X' in the appropriate box(es)*

☐ A. Air ☐ B. Rail ☐ C. Highway ☐ D. Water ☐ E. Other *(specify)*

IX. First or Subsequent Notification

Mark 'X' in the appropriate box to indicate whether this is your installation's first notification of hazardous waste activity or a subsequent notification. If this is not your first notification, enter your installation's EPA ID Number in the space provided below.

☐ A. First Notification ☐ B. Subsequent Notification *(complete item C)*

C. Installation's EPA ID Number

EPA Form 8700-12 (Rev. 11-85) Previous edition is obsolete.

Continue on reverse

Figure 11.1

ID — For Official Use Only		
C		T/A C
W		1

X. Description of Hazardous Wastes *(continued from front)*

A. Hazardous Wastes from Nonspecific Sources. Enter the four-digit number from 40 *CFR* Part 261.31 for each listed hazardous waste from nonspecific sources your installation handles. Use additional sheets if necessary.

1	2	3	4	5	6
7	8	9	10	11	12

B. Hazardous Wastes from Specific Sources. Enter the four-digit number from 40 *CFR* Part 261.32 for each listed hazardous waste from specific sources your installation handles. Use additional sheets if necessary.

13	14	15	16	17	18
19	20	21	22	23	24
25	26	27	28	29	30

C. Commercial Chemical Product Hazardous Wastes. Enter the four-digit number from 40 *CFR* Part 261.33 for each chemical substance your installation handles which may be a hazardous waste. Use additional sheets if necessary.

31	32	33	34	35	36
37	38	39	40	41	42
43	44	45	46	47	48

D. Listed Infectious Wastes. Enter the four-digit number from 40 *CFR* Part 261.34 for each hazardous waste from hospitals, veterinary hospitals, or medical and research laboratories your installation handles. Use additional sheets if necessary.

49	50	51	52	53	54

E. Characteristics of Nonlisted Hazardous Wastes. Mark 'X' in the boxes corresponding to the characteristics of nonlisted hazardous wastes your installation handles. *(See 40* CFR *Parts 261.21 — 261.24)*

☐ 1. Ignitable *(D001)* ☐ 2. Corrosive *(D002)* ☐ 3. Reactive *(D003)* ☐ 4. Toxic *(D000)*

XI. Certification

I certify under penalty of law that I have personally examined and am familiar with the information submitted in this and all attached documents, and that based on my inquiry of those individuals immediately responsible for obtaining the information, I believe that the submitted information is true, accurate, and complete. I am aware that there are significant penalties for submitting false information, including the possibility of fine and imprisonment.

Signature	Name and Official Title *(type or print)*	Date Signed

EPA Form 8700-12 (Rev. 11-85) Reverse

Figure 11.1 *(cont.)*

classified according to its principal component. Contaminated flammable solvents can generally be commingled, as can various waste chlorinated solvents. However, waste flammable solvents and waste chlorinated solvents should not be commingled.

Used motor oils and lubricants may be commingled, but gasoline, benzene, kerosene, and other solvents should not be mixed in with the oils. Waste acids and alkalis should not be mixed, except under the direction and guidance of a qualified chemist.

Routinely, all waste streams should be segregated from each other and should be assumed incompatible until proven otherwise.

Small Quantity Generator

Federal law currently defines a *small quantity generator* as anyone who generates less than 2,200 lbs. of hazardous waste per month, or less than 2.2 lbs. of acutely hazardous waste per month. State laws vary on this point. Figure 11.2 summarizes state laws which define small quantity generators in terms other than those used by the federal government.

Small quantity generators are allowed to store the hazardous wastes they generate for up to 180 days under federal law. Again, state laws vary. Figure 11.2 also shows the maximum number of storage days allowed by those states whose regulations differ from the federal standard.

Large Quantity Generator

Federal and state laws generally define a *large quantity generator* as any person generating hazardous wastes in quantities greater than those which define a small quantity generator. Both the federal government and the various states specify shorter storage periods for large quantity generators than for small quantity generators. Figure 11.2 summarizes those state laws which define large quantity generators in terms other than those used in RCRA. Figure 11.2 also shows the maximum storage times allowed for large quantity generators by the various states which require storage times different from those provided in RCRA.

Appendix C provides a list of the various state hazardous waste management agencies. Those officials should be contacted for detailed information regarding storage requirements and generator status related to specific wastes.

Liquid Waste Storage

Most liquid wastes are stored in 35-gallon or 55-gallon metal drums. These drums are readily available, easily moved when full, and provide spill-proof seals when properly fitted with screw-in bungs. Filling is accomplished either by direct pumping or by pouring liquids from transfer containers through a funnel.

Except for strongly acidic wastes, such as used battery acids, use of metal drums for liquid waste storage is recommended. Battery acids should be stored in compatible rigid plastic containers, such as low-density polyethylene, high-density polyethylene, polypropylene, or polymethylpentene (TPX).

Storage areas should be designed such that a catastrophic failure of all containers in the area will be fully contained. Such precautions usually require a diked berm to be built around the storage site. Moreover, the area must be maintained at a temperature above 35°F to prevent freezing liquids from rupturing the storage drums, and below 120°F to prevent excess pressure build-up inside the drums, which could also lead to rupture. Note that federal regulations require only that 10% of the total contained volume, or 110% of the volume of the largest container in the storage area, must be

Summary of State Variations Regarding Small/Large Generators	
State	**Comments**
California	No small quantity generator exclusion, except for generators of less than 100 kg (220 lbs)/month. 90-day storage limit.
Connecticut	State regulations allow maximum of 180 days of on-site storage of up to 1,000 kg (2,200 lbs). No extension to 270 days storage is allowed for shipments of over 200 miles.
District of Columbia	Generators of more than 50 kg/month are considered "large generators," with 90-day storage limit.
Florida	Generally the same as EPA rules, but no 270 days/200 miles rule.
Kansas	25 kg/month limit for "conditionally exempt small quantity generators." Small quantity generators are those producing 25-1,000 kg/month.
Kentucky	Generators in 100-1,000 kg/month class treated same as large generators, i.e., 90-day storage limit unless 270 days/200 miles rule applies.
Louisiana	No automatic exclusion – must be applied for by the generator.
Maine	2 categories: small quantity exemption for less than 100 kg/month. If over 100 kg/month, 90-day storage limit.
Maryland	All generators above 100 kg treated similarly.
Massachusetts	Essentially same as EPA rules, but no 270 days/200 miles rule.
Montana	Same three categories of generators, but large generators (over 1,000 kg/month) are treated the same as 100-1,000 kg/month generators. EPA has objected to state's rules, so 90-day storage limit should be assumed.
Nebraska	No 270 days/200 miles rule.
New Jersey	Small generator threshold is 100 kg/month, so 90-day storage limit.
New Hampshire	Small generator threshold is 100 kg/month, so 90-day storage limit.
New York	Small quantity generator threshold is 100 kg/month. If over 100 kg/month, must store less than 1,000 kg to be exempt from some requirements. If storing more than 1,000 kg, must store only 90 days.
North Dakota	All 100-1,000 kg/month generators can store for 270 days because all wastes in state must be shipped over 200 miles.
Oregon	Small quantity disposal exemptions are assigned individually to specific wastes – check with Oregon Dept. of Environmental Quality.
Pennsylvania	Generally same as EPA rules, but storage on-site of over 1,000 kg causes small quantity generator to be treated as large generator with 90-day storage limit.
Rhode Island	No exemptions – 90-day storage limit.
South Carolina	If over 100 kg/month, 90-day storage limit.
Vermont	Generators of more than 100 kg/month may store for 180 days up to 1,000 kg if there are proper storage facilities.

Figure 11.2

containable within the diked area. The more stringent 100% total volume containment is recommended, however, wherever possible.

To minimize the potential for mixing incompatible wastes, it is recommended that storage areas be color-coded for the various wastes being stored. For example, red is for acids, blue is for alkalis, green is for motor oils, yellow is for paint wastes, etc. The employee Right-to-Know training session and Hazards Communication Manual should clearly spell out the color coding system. Storage areas for incompatible materials must be separated by berms, dikes, or walls.

Figure 11.3 depicts a standard storage shed for drums of used motor oils, solvents, paint wastes, or other liquids. The shed as drawn is suitable for a small quantity generator; a similar structure is recommended for a large quantity generator.

Solid Waste Storage

Hazardous solid wastes, other than asbestos wastes, which are discussed in the next section of this chapter, are usually composed of damp absorbents used to absorb spills of hazardous liquids, the rags used to apply or clean up solvents, dried paints, and the containers that once held hazardous materials. These items are best stored in metal or plastic drums with covers that open completely to facilitate filling and dumping.

Dry materials, such as plastic sheeting used to contain contaminated soils, and contaminated personal protective equipment, can be stored in paper drums with tightly fitted lids. Materials containing no free liquids do not require a bermed storage area, provided that (a) the actual area is sloped to prevent water accumulation and (b) the dry materials are raised above the floor slightly, as on pallets. The use of conventional dumpsters for hazardous waste storage is discouraged due to the high probability of either inadvertent disposal as a non-hazardous waste, or the disposal of hazardous waste to the wrong dumpster. In addition, cleaning a dumpster after using it for hazardous waste storage is a risky undertaking.

Solid wastes are usually not harmed by exposure to freezing. High temperatures (greater than 120°F) can, however, cause spontaneous combustion. Exposure to the elements, particularly in the case of paper drums, can lead to rapid drum deterioration and leakage.

The design of storage facilities for solid hazardous wastes should emphasize protection from the elements and protection from damage to the containers. Figure 11.4 depicts a suitable storage facility for solid hazardous wastes. Note the requirements for ventilation to prevent excessive heat build-up.

Automotive parts can create special storage and disposal problems. Used oil filters, automotive batteries, brake shoes, clutch pads, etc., pose special problems. They should not be routinely mixed with other hazardous solid wastes.

Used oil filters should be carefully drained to a waste oil container before disposal. Since EPA, along with most state regulatory agencies, does not consider waste oil to be a hazardous waste, disposal of drained oil filters with regular construction debris is acceptable. For states that do consider waste oil as hazardous, used oil filters will have to be treated as a hazardous solid waste.

Waste automotive batteries contain a strong acid. However, since they are recyclable with the acid still in them, no attempt should be made to empty the battery. Most batteries currently available are sealed, and any attempt to drain them will usually lead to an acid spill.

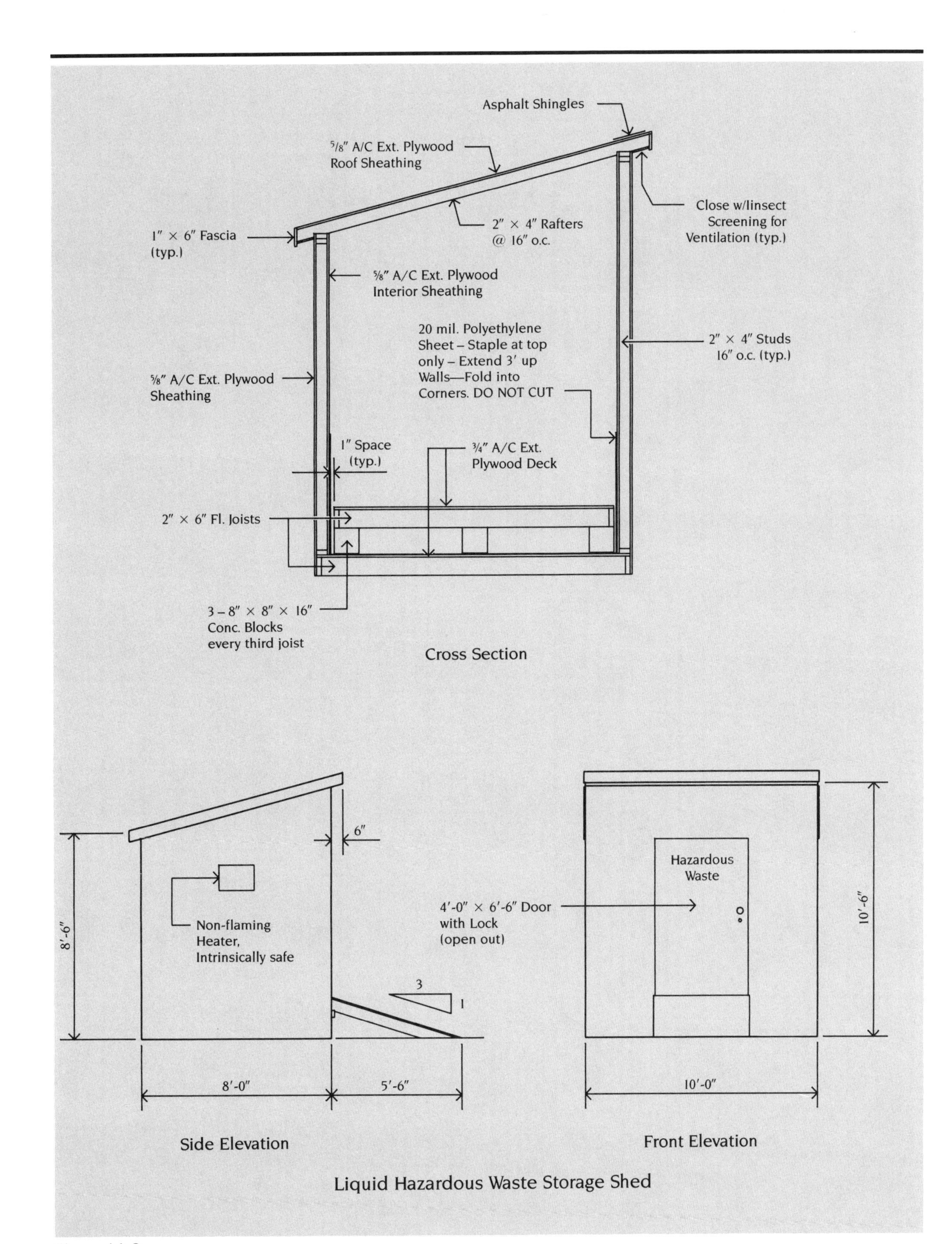

Asphalt Shingles
5/8" A/C Ext. Plywood
Roof Sheathing
Close w/Insect
Screening for
Ventilation (typ.)
1" × 6" Fascia
(typ.)
2" × 4" Rafters
@ 16" o.c.
5/8" A/C Ext. Plywood
Interior Sheathing
20 mil. Polyethylene
Sheet – Staple at top
only – Extend 3' up
Walls—Fold into
Corners. DO NOT CUT
2" × 4" Studs
16" o.c. (typ.)
5/8" A/C Ext. Plywood
Sheathing
1" Space
(typ.)
3/4" A/C Ext.
Plywood Deck
2" × 6" Fl. Joists
3 – 8" × 8" × 16"
Conc. Blocks
every third joist
Cross Section
6"
8'-6"
Non-flaming
Heater,
Intrinsically safe
3
1
8'-0"
5'-6"
Side Elevation
Hazardous
Waste
4'-0" × 6'-6" Door
with Lock
(open out)
10'-6"
10'-0"
Front Elevation
Liquid Hazardous Waste Storage Shed

Figure 11.3

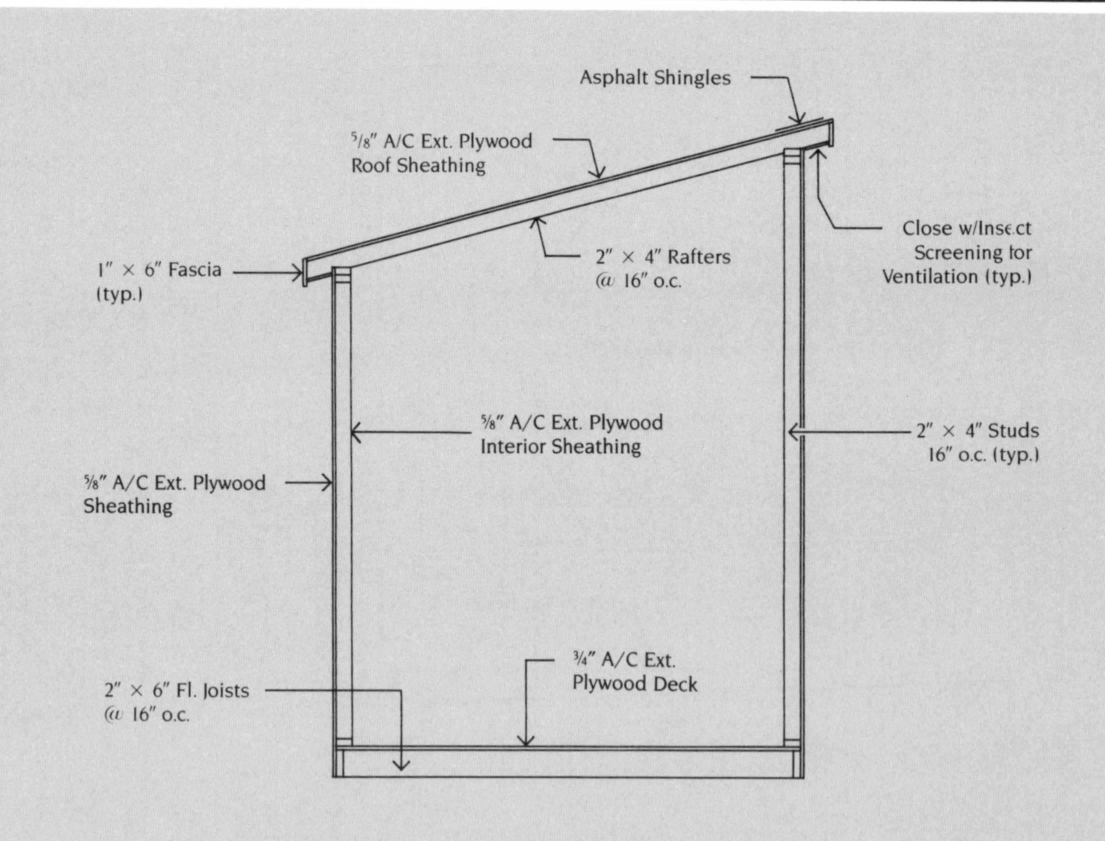

Cross Section

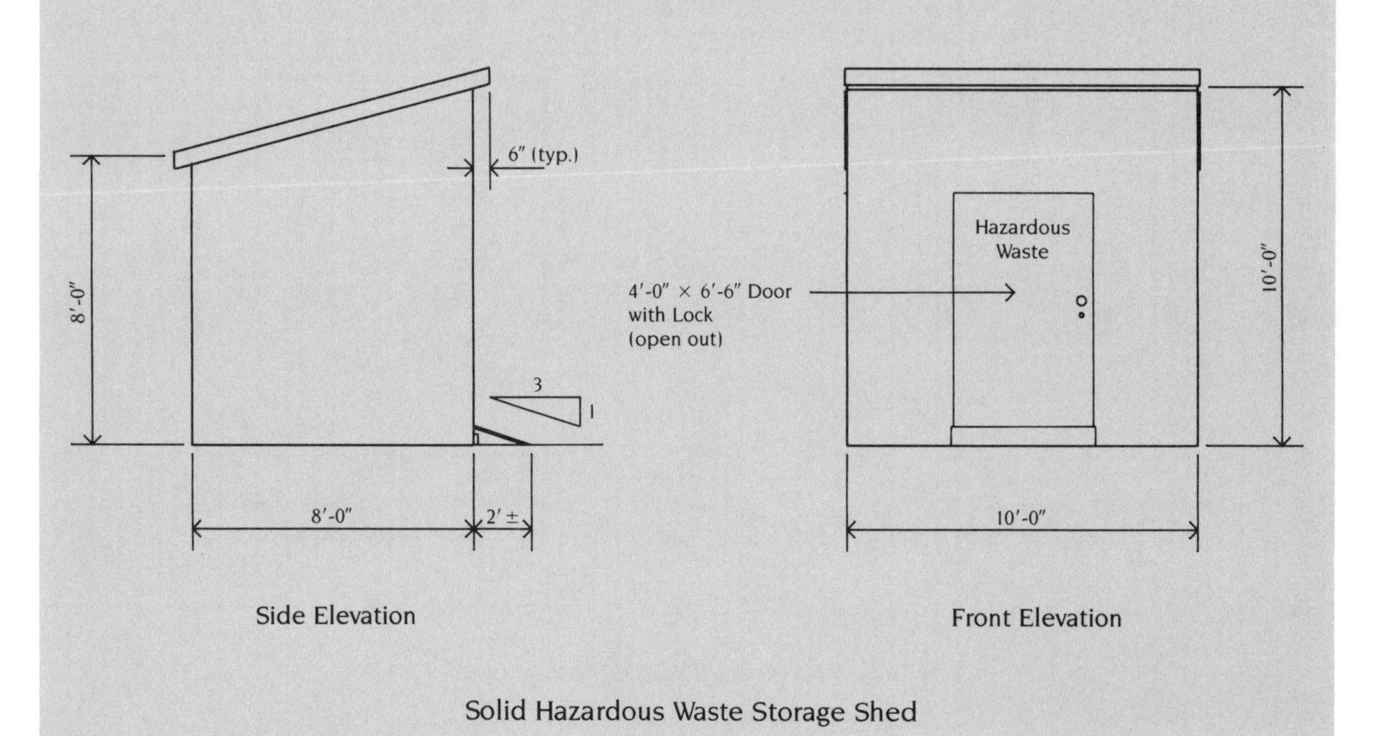

Side Elevation

Front Elevation

Solid Hazardous Waste Storage Shed

Figure 11.4

Waste batteries should be stored on a wood pallet, away from alkalis and other incompatibles. Storage in an area designed to prevent freezing and/or accidental damage is recommended. Storage sheds used for liquid acid wastes are also suitable for waste batteries.

Brake shoes and clutch pads often contain asbestos. Those that do must be disposed as a hazardous waste in accordance with local asbestos disposal regulations. In general, asbestos wastes require double bagging in plastic bags and disposal at a special landfill. See Chapter 13 for a further discussion of asbestos waste handling and disposal.

Labeling

Current law requires the proper labeling of all containers used to store hazardous wastes. Storage sheds must be similarly marked.

Figure 11.5 depicts a standard hazardous waste label suitable for attaching to storage drums and buildings. These labels are commercially available in a pre-printed format for easy use. Appendix D provides guidance for determining the Department of Transportation (D.O.T.) shipping names, hazard class, and UN or NA number, for some of the more common types of wastes found on a construction site. For other wastes, check the Material Safety Data Sheet supplied with the principal constituent, and contact the suppliers of that material. For a complete list of all materials assigned a D.O.T. shipping name and UN or NA number, consult current federal D.O.T. regulations in the Code of Federal Regulations (CFR), 49 CFR 172.101.

HAZARDOUS WASTE

FEDERAL LAW PROHIBITS IMPROPER DISPOSAL
IF FOUND, CONTACT THE NEAREST POLICE
OR PUBLIC SAFETY AUTHORITY, OR THE
U.S. ENVIRONMENTAL PROTECTION AGENCY

MANIFEST # ____________

D.O.T. SHIPPING NAME ____________

UN OR NA # ____________

EPA NAME ____________

EPA # ____________

GENERATOR INFORMATION:

NAME ____________

ADDRESS ____________

CITY ____________ STATE ____________ ZIP ____________

DATE OF
GENERATION/ACCUMULATION ____________

HANDLE WITH CARE!

Figure 11.5

Chapter 12

Waste Minimization and Reduction

Hazardous waste storage and disposal costs tend to be high. It pays, therefore, to carefully consider waste minimization and reduction strategies as a part of the overall waste management plan. This chapter outlines specific strategies that can be used to accomplish those goals. Strategies discussed include inventory control, in-house recycling, segregation of waste streams, and minimization of material usage.

Inventory Control

Before one can develop a strategy to reduce wastes, one must know the materials that are being used which could generate wastes, and the resulting waste streams. Inventory control is used to generate this information.

Chapter 9 provides a discussion of the management planning necessary to ensure the safe storage and handling of hazardous materials and hazardous wastes. Part of that planning effort includes the preparation of a material inventory, by work area, in accordance with the requirements of the Right-to-Know law. Figure 9.2 shows an inventory sheet used to accumulate this data.

By alphabetizing the inventory sheets by work area, and then reviewing the use column for each inventory, it should be easy to spot the existence of multiple materials used for the same purpose. By then comparing work area inventories, it should not be difficult to identify multiple materials used for the same purpose in different work areas.

For example, the workers building a patio may be using two different brands of brick cleaner, while those building a fireplace nearby may be using still a third brand. This approach makes little sense unless there is a specific reason for using each cleaner at each job site. The task of management is to identify the situation and to either verify the need for three different brands, or identify that brand least likely to cause environmental or health risks if improperly used or spilled. If one of the brands will serve all of the needs with less risk, then the extraneous brands should be eliminated.

Having decided on the brand of material to use, it is important to identify the total quantities needed and the timing of the need. It makes no sense to order five drums of brick cleaner early in the project if the only brick work involves hidden chimneys and a small patio. Moreover, if a 5-gallon pail will be sufficient, a 35-gallon drum will only generate unnecessary hazardous waste and a high potential for spillage.

The keys to inventory control, then, are:

1. To carefully review inventory sheets to eliminate duplication
2. To monitor ordered quantities and the time when the materials are needed

In-House Recycling

One of the best ways to minimize waste is to convert a substance into a raw material. For example, it is common for firms to use a variety of cleansers for different jobs, since each job requires a different degree of cleanliness. Even when the various cleansers are grouped together so that only one type gets used for all jobs, a lot of material still tends to be wasted unnecessarily. In-house recycling is a way to control the waste.

Specifically, degreasing solutions used to dip-clean miscellaneous metal parts will, at some point, become too dirty to be effective for this purpose. However, these moderately dirty solutions can still be used in the motor pool area to clean the dirt and grease off engine parts very effectively. This eliminates the need to dispose of the cleaner as a waste, initially, and eliminates the need to buy additional degreaser for the motor pool. After the motor pool has finished using the solvent, it may be possible to screen out the dirt and use the liquid as a fuel for smudge pots or for roofing tar heaters. Again, the waste is minimized and raw material purchases are reduced.

Waste Reduction

The concept of waste reduction encompasses more than simply recycling wastes and avoiding duplication of materials. It also means considering what is in the various waste streams and where each one goes.

It is very easy to throw all refuse into a large metal dumpster and let the hauling contractor worry about what to do with it. Unfortunately, that practice regularly leads to soil and groundwater contamination. When the disposed waste can be traced back to a particular generator, that generator can be held liable for the total cost of cleaning up the soil and water. To avoid those costs, it is necessary to dispose of hazardous waste separately in approved, licensed hazardous waste disposal facilities. These precautions can be expensive, but not nearly as expensive as groundwater cleanup.

The cost of hazardous waste disposal differs depending on the substance. Asbestos and PCB's, for example, are very expensive to dispose because the disposal facilities are heavily regulated and only a few exist. Waste motor oil, on the other hand, is fairly inexpensive to dispose because it is easily converted to a useful fuel for low-quality heat generators.

If one gallon of PCB-contaminated oil is mixed with 1,000 gallons of waste motor oil, the entire amount, all 1,001 gallons, requires PCB disposal techniques rather than waste motor oil techniques. Thus, what was once a somewhat costly one-gallon disposal problem suddenly becomes a very expensive 1,001-gallon disposal problem.

Waste reduction requires a manager to monitor the waste streams to ensure that only compatible wastes are mixed. Mixing of incompatible wastes, from both a chemical reactivity standpoint and an end-disposal standpoint, can be very costly.

Education

Management can, theoretically, prepare perfect inventories, order limited quantities of specific materials, and plan the "daylights" out of a project. But if the employees do not follow the plan, all of management's efforts will be for naught. Education of the employee is critical to the success of waste minimization and reduction efforts.

As outlined in Chapter 3, the Right-to-Know laws require an employer to provide all employees with specific training on the health hazards found in the employee's work area. That training session is also the perfect place to explain the rationale and intent of the firm's waste minimization and reduction plans. When workers understand *why* the manager wants used degreasers taken to the motor pool, they are much more likely to follow the plan. Employers should not be shy about explaining the "why's" behind hazardous waste management plans, along with the "how-to" directives.

Chapter 13

Asbestos and PCB Wastes

Asbestos and PCB wastes are usually generated on renovation projects rather than in the course of any new work. That is because PCB's are now banned from new construction use, and asbestos is nearly banned. PCB's are still used as a component in some materials, particularly in electrical system component lubricants. Asbestos persists in older buildings in the form of asbestos shingles, pipe insulation, and fireproofing.

These materials should only be handled by experts, as noted in Chapter 7. Nevertheless, it is recognized that occasionally a small quantity of asbestos or PCB wastes will find its way to some job sites. It is important that contractors and their employees know how to deal with these situations. This chapter will address that issue.

Removing Asbestos Debris

Since asbestos is only hazardous when airborne, and since it only becomes airborne from a friable, crumbly state, the key to the safe cleanup of asbestos wastes is to keep the material from becoming friable. This can be accomplished either by encapsulating the particles with a cementitious binder, or wetting the product sufficiently to cause the particles to either stick together or become so heavy as to prevent air flotation.

A variety of encapsulating products are currently available on the commercial market. All of these products tend to be absorbed into the asbestos before setting up, thereby ensuring the long-term stability of the asbestos. The problem with these products for cleanup projects is that the asbestos particles are usually scattered widely in an area and the encapsulating product merely serves to bridge the gaps between the particles. In so doing, it adheres to everything else in sight and glues the debris down rather than allowing it to be cleaned up. These characteristics make encapsulation products unsuitable for many cleanup projects.

Encapsulants can be used, and in fact are very effective for, sealing the exposed edges of damaged asbestos. This procedure should only be done, however, by skilled workers.

There is only one sound method to use. That is to wet the asbestos debris, thereby reducing the opportunity for particles to become airborne.

Asbestos Cleanup Procedure

Before starting the cleanup, evacuate the area of all nonessential personnel. Make sure that the cleanup crew has donned HEPA-filtering face masks and disposable protective clothing. Clothing should include head covering (worn over the face mask), coveralls, booties, and gloves. The head covering, booties, and gloves should be sealed to the coveralls. Only properly trained employees should conduct the cleanup. If none are available in-house, seal the area and call in a trained cleanup contractor.

Isolate the work area. If a room can be isolated by merely closing doors and windows tightly, do so. Otherwise, seal the space with 6-mil plastic sheeting duct-taped to walls, ceilings and floors to completely enclose the contaminated area. Place asbestos warning signs on the doors or plastic sheeting. Make sure all vents and ducts are also sealed.

Next, gather the tools needed to perform the cleanup. Figure 13.1 lists the equipment required. If the debris is dry or damp, but not wet, and small in size, it can be vacuumed up with an HEPA-filtered vacuum. A regular vacuum must not be used because the bag will not trap the asbestos fibers, which will be blown around the room, exacerbating the problem.

Required Equipment for Small Asbestos Debris Cleanup

	Item	Cost
1.	Large garden sprayer or small plant mister filled with amended water*	$ ________
2.	6 mil, pre-labeled, polyethylene storage bags	________
3.	Asbestos hazard or warning labels	________
4.	HEPA-filtered vacuum cleaner	________
5.	Mop and bucket filled with amended water*	________
6.	Rags	________
7.	Duct tape	________
8.	Shovel, dustpan, or garden trowel	________
9.	Ice scraper or another dustpan	________
10.	Half-face reusable respirator with NIOSH-approved cartridges and filters	________
11.	Protective disposable clothing	________
12.	One roll of 12′ x 100′ polyethylene sheeting (4 or 6 mil)	________
13.	Zip-lock freezer bags large enough to hold respirator	________
	Total Cost	$ ________

**Amended water is clean tap water to which a surfactant has been added. The recommended surfactant (a detergent) consists of 50% polyoxyethylene and 50% polyoxyethylene ether mixture. One ounce of this material, or equivalent surfactant, is added to 5 gallons of water to create an amended water.*

Figure 13.1

After vacuuming thoroughly, wet-mop all surfaces upon which the debris has fallen. Walls, ceilings, windowsills, shelves, and moldings should all be wiped carefully with a damp cloth rag. All visible debris must be wiped or mopped up.

If the debris is too big or too wet to be vacuumed, it should be thoroughly wetted using a light mist. Slowly wetting the debris with a mister will help to avoid stirring up fiber dust. A hard spray should not be used, as it will put too much water down too quickly, allowing fibers to spread on the puddles of water, and raising other fibers into the air.

After the debris is thoroughly wetted, pick up larger pieces and carefully place them into a pre-labeled 6-mil polyethylene bag. Pick up smaller pieces with a shovel, dust pan, or trowel and put them into the same or similar bags. Use an ice scraper, trowel, or dustpan to push material onto the shovel for pickup. Do not use a broom or brush. Brooms and brushes are difficult to clean and generate a much higher likelihood or fiber flotation.

Carefully wash all shovels, trowels, scrapers, and other tools that were used to collect asbestos waste. Hold the items over a pre-labeled plastic bag and gently spray with the garden sprayer or mister. If any item will not wash clean, or is not washable, throw it away. Put it in the bag with the debris. Do not try to keep it, and do not remove it from the cleanup area except inside the sealed debris bags.

Then wet-mop the entire area and clean all surfaces with a wet rag. Note that if the area is carpeted, the carpet must be steam cleaned (in-place). A HEPA-filtered vacuum cannot be used to pick up water or wet material.

The bags containing debris and waste tools should be twisted shut and sealed with duct tapc. The sealed bags are then placed inside a second 6-mil bag, and that bag is sealed the same way. After all the debris has been bagged and sealed and the area has been wet mopped and wiped clean, all mopheads, rags, tools that cannot be cleaned, and protective clothing are placed inside a separate bag, which is sealed as above. That bag, too, is then sealed inside a second bag.

Used respirators are sealed in zip-locked bags for cleaning and disinfection by certified technicians or disposed in the same manner as the other protective clothing.

Dirty water in the bucket is poured down a sink—not onto the ground surface. The bucket and sink must then be thoroughly washed clean.

After the initial cleanup is complete, the area must remain isolated until it is completely dry. A careful inspection is then made by a person wearing a protective respirator. Sometimes wet asbestos debris will remain hidden during a cleanup operation. It will become visible again after it dries. If any such material is found during the inspection, the wet mop, HEPA-filtered vacuum, or steam-cleaning operation must be repeated.

When the area has been determined to be safe, the barriers are carefully removed and sealed in pre-labeled 6-mil polyethylene bags. Those bags are then sealed inside a second set of bags. All the bags are then disposed as a manifested waste through a licensed disposal contractor.

Outdoor Asbestos Debris

Any asbestos debris found outside a building should be quickly and carefully covered with a thin (one-or two-inch) layer of soil or sand. The area should then be isolated with stakes and survey tape. A cleanup contractor should then be called to encapsulate the material in place and remove it for disposal at a licensed disposal facility. Outside debris must not be allowed to blow around and cleanup should not be attempted by untrained or unlicensed personnel.

Asbestos Warnings

Dealing with asbestos waste is a particularly hazardous business. The procedures outlined herein are cumbersome, time-consuming, and expensive. Nevertheless, the risks are too great to "cut corners." People who do not respect the dangers of asbestos die early, painful deaths.

Note, too, that most states and municipalities have enacted laws or issued regulations to control the cleanup of asbestos. Before any cleanup is attempted, the local Board of Health should be contacted to ensure compliance with local variations on the procedures set forth above.

It is the opinion of the authors that no contractor should attempt any asbestos cleanup project unless the personnel employed have been thoroughly trained to perform the work. The procedures outlined in the previous section are provided to assist those who resist the admonition to call in an expert. These procedures relate only to the cleanup of very small quantities of debris from inside a work area. Even those jobs are best left to experts.

The Source of PCB Wastes

PCB wastes are almost always associated with the disposal of old electrical system components, particularly transformers. The PCB's were used as admixtures to lubricating oils due to the various insulating properties that they impart to the oils. Those oils are usually removed with the electrical equipment. They remain inside until the equipment is opened at a licensed disposal facility and the oil is properly disposed. That is clearly the best method of handling PCB wastes.

PCB Spills

Occasionally, an electrical component will break or spill, and wastes containing PCB's will contaminate an area. In such cases health risks are associated with ingestion, rather than inhalation. That means that a spill of PCB waste is not inherently dangerous, as it exists on the floor or the ground. Cleanup activities are designed to minimize the spread of the material, to avoid direct skin contact by people, and to dispose of the wastes in a safe manner.

The first action in the event of a spill should be to stop the flow, using a dry absorbent or sand to soak up the oil. If the PCB spill is on soil, it should be absorbed therein, without the use of additional absorbent. Care should be taken, however, to avoid overland run-off of large spills or contamination of water bodies. In wet weather, an outdoor spill should be quickly stabilized with absorbent and covered with 6-mil polyethylene sheeting to prevent contamination of groundwater.

Indoor spills should be kept away from floor drains, sinks and sumps. Sink disposal is not an appropriate disposal technique for PCB's.

PCB Removal

If workers come in direct contact with PCB-bearing waste, the affected skin areas should be thoroughly washed with a mild solvent (such as alcohol or grease-cutting hand cleaners), then washed with soap and water to remove the solvent. Contaminated clothing should be immediately removed and disposed in a 6-mil polyethylene bag, appropriately labeled. Cleanup crews should don protective coveralls, booties, and gloves, taped to the coveralls. They should also wear dust masks and goggles before attempting any cleanup activities.

PCB *Cleanup*

Large spills, whether indoors or out, should be cleaned up only by trained contractors. However, a small spill can be safely cleaned up by workers on the site if they adhere strictly to the following procedures.

First, gather together the required tools and equipment. (Figure 13.2 lists the equipment necessary for a small PCB waste cleanup.) Then isolate the area with signs and barriers and ensure that cleanup personnel are wearing proper protective equipment.

Next, carefully shovel the absorbent material or contaminated soil into 30-mil plastic bags. Do not overfill. Twenty-five to thirty pounds is enough for any one bag. Seal each bag by twisting the top closed and sealing it with duct tape. Seal those bags inside a second set of bags in the same manner. Use an ice scraper, piece of wood, or other disposable object to push the material onto the shovel. Dispose of the pusher in the bag with the contaminated wastes.

Required Equipment for Small PCB Spill Cleanup

	Item	Cost
1.	6 mil pre-labeled, polyethylene storage bags	$ ________
2.	PCB hazard or warning labels	________
3.	Dust mask* or respirator*	________
4.	Coveralls, booties, gloves, goggles	________
5.	Duct tape	________
6.	Pail or bucket with hot, soapy water**	________
7.	Appropriate solvent for PCB carrier	________
8.	Dry absorbent	________
9.	Shovel, dustpan, trowel	________
10.	Ice scraper	________
11.	Rags	________
	Total Cost	$ ________

**Check solvent* MSDS *for respirator requirements.*
***Soap must be compatible with solvent used to remove* PCB *carrier.*

Figure 13.2

Outdoors, dig the hole big enough to ensure that all contaminated soil is in the bag. Indoors, or on concrete or other non-disposable outdoor surfaces, further cleanup of residuals is required. Non-disposable surfaces must be washed with an oil solvent. Follow the MSDS recommendations for handling the solvent being used. Wipe up the spill with solvent-dampened rags and deposit the rags into a 6-mil polyethylene bag. Remember that the solvent is a hazardous waste itself, and that it will be contaminated with the PCB's during cleanup. The cleaning materials must be properly disposed with the PCB wastes.

If the solvent used leaves a residue, the residue must be cleaned with soap and water. Care should be exercised to avoid washing solvent-contaminated soapy water into the ground when working outdoors. All soap, water, and solvents should be absorbed using rags which are put into the 6-mil bags for ultimate disposal.

When the wastes have been cleaned up, the protective clothing and dust masks are carefully removed and placed in a 6-mil polyethylene bag. The bag is sealed as above and then double-bagged, as above. All bags are then disposed through a licensed hazardous waste disposal contractor.

PCB *Warnings*

All spills of PCB's are best left to the experts. The procedures outlined above are provided only because some people will always try to do it themselves. In that case, it is important to at least follow a safe procedure.

The PCB cleanup process outlined in this chapter admittedly is cumbersome. People will be tempted to skip the dust mask and protective clothing, to dump the absorbent or soil into barrels or boxes, and to otherwise cut corners. However, the risks are tremendous, and the time-saving shortcuts are not worth the costs. Those who tempt fate lose.

Finally, note that PCB's do not break down quickly over time. Merely burying a small spill in the back yard will not make it disappear. This approach will, in fact, create a long-term liability that could erupt at any time. PCB's are hazardous materials, and there is good reason for the restrictions that have been assigned to them. PCB's that are properly handled will not cause serious harm to people.

Chapter 14

Unexpected Subsurface Contamination

One of the biggest and most expensive surprises on a construction site is the discovery of hidden, subsurface contamination. After months of careful bidding, planning, and organizing, the contractor mobilizes on the site, ready to go to work, and the first thing his excavator encounters is a soil of unusual color under the overlying gravel, a strong odor from the hole, or a pile of leaking drums buried under the surface. The contractor has no other choice than to stop the work until the nature and extent of the problem can be ascertained and a cleanup procedure developed.

The Problem

Contamination of a site can occur in one of several ways. Direct disposal of a hazardous substance to the ground surface is the most common. In addition, however, subsurface injection of materials, direct burial of containers, and subsurface infusion of contamination through natural groundwater flow from up-stream contaminated sites could also cause contamination of the subject parcel. Each of these routes must be checked by conducting a site evaluation whenever evidence of contamination is found.

The Solution

Two steps are required to evalute the contamination potential from direct disposal of substances to the ground surface. The first involves a check of historical town records to determine prior site uses. Such information often suggests possible locations of dumping or burial on the site, and the nature of contaminants that might be found. The second step involves physically walking the site, particularly in those areas suggested by the records search, to see if there are any visual indications of prior disposal, such as stressed or missing vegetation, localized soil discoloration, regular-shaped depressions, or odors. Any areas identified as likely sites of contamination require excavation of a test pit and/or installation of a groundwater sampling well.

Most contaminants tend to dissipate within a certain amount of time following a disposal activity. Dissipation occurs through volatilization to the atmosphere and through washing of the contaminant into the soil by rain or melted snow. Contaminants that are washed off the surface typically end up in surface ponds or streams. When those areas are stagnant or marshy, the contaminants tend to collect and remain for long periods of time. When contaminants wash into the soil, they collect in the groundwater under the site and spread out in a "plume" with the natural groundwater flow. Sampling of surface water and groundwater is necessary to verify these types of contamination.

Another way in which contamination may enter a site is through groundwater flow from an adjacent parcel of land. The best way to identify this source of contamination, as well as any future incursions, is to set a series of groundwater monitoring wells along the up-stream property line. The wells are then sampled on a regular basis (every year or two) for various indicator pollutants.

Testing

Testing of water samples should initially be performed only for the purpose of identifying *indicator pollutants*. These are substances that can be easily measured and indicate the presence of one or more pollutants, but are not necessarily harmful. The indicators normally suggested are volatile organic compounds (VOC's), electrical conductivity, and pH. Depending on the nature of the contaminants found or suspected, however, additional tests may be indicated.

The VOC test is an indicator of volatile contaminants, such as gasoline, benzene, toluene, and other organic compounds. Organics are those compounds which contain carbon, hydrogen, and/or oxygen molecules. These organics which are troublesome contaminants are often volatile as well.

The electrical conductivity test is an indicator of metal ions in the water. Whenever metals dissolve in water, they ionize to varying degrees. The increase in ions changes the electrical conductivity of the water, indicating contamination. Metals include sodium, copper, iron, manganese, magnesium, and other, so-called "heavy" metals.

The pH test is less indicative of a specific problem. pH is the relative acidity or basicity of a sample. Changes in the pH, together with changes in the other indicators, often can be used to suggest a cause for other irregular test results. The pH test is a useful, inexpensive diagnostic tool.

Both the pH and conductivity tests are done at the well or source during sample collection. The VOC test can also be conducted at the source, using a portable analyzer. A more accurate VOC test can be performed in a laboratory. For monitoring purposes, however, field testing should be adequate. When a significant concentration is indicated in field results, samples for a laboratory testing should be collected and tested to confirm and refine the results.

Finalizing the Evaluation

The program outlined involves a records search; a site walkover; the installation of groundwater monitoring wells; and field testing of the new wells, any existing wells, and existing on-site surface waters. When all the data has been collected, a comprehensive written report is prepared outlining the study findings, conclusions, and recommendations.

Cleanup

A cleanup program is developed from the site evaluation data. It may be possible to process soil contaminated with gasoline, oil, or organic solvent using a mobile treatment unit. That unit will heat the soil and contaminant to levels high enough to evaporate the contaminant, and then burn the gases in a secondary combustion chamber. The cleaned soil can then be returned to the site.

Other soil treatment options many include in-situ biological treatment, in-situ vitrification, air stripping, steam stripping, removal to a secure landfill, removal to an on-site landfill constructed for that purpose, and other methods devised on a case-by-case basis.

If a PCB or other dangerous material is present, a very high temperature mobile burner may be used. The mobile burner burns the whole mass at a temperature high enough to destroy the contaminants. It also allows the cleaned soil to be returned to the site.

Goundwater contamination may require extensive pumping and treatment strategies. Once the surface soils are cleaned, however, groundwater cleanup will probably not delay the construction project. It should be noted that any site dewatering activities must be coordinated with the treatment scheme in order to avoid spreading contaminants.

Drums, barrels, and other solid contaminant sources can often be stabilized in place, then removed intact for disposal at a licensed diposal facility. The appropriate stabilization process and disposal costs are contaminant-specific.

In all cases, it is very unlikely that the general contractor will have the skilled employees on-hand, trained to perform a cleanup of subsurface contaminants. The contractor's only course of action is to stop work immediately upon discovery of such hazards, evacuate the area, and call in a cleanup specialist to resolve the problem.

Chapter 15

Leaking Underground Storage Tanks

The identification and removal of leaking underground storage tanks (LUST) are subjects that have begun to receive a lot of attention. The potential for adverse environmental and human health impacts from even a single leaking tank is enormous. The risk associated with removal of those tanks, however, is equally enormous. This chapter presents a series of protocols designed to help engineers and contractors identify and deal with the risks inherent in LUST management.

Site Identification

Leaking underground storage tanks are identified through three general mechanisms.

- Regular testing of the tanks (which suddenly begin to show an otherwise inexplicable loss of product)
- Excavation of a pit in preparation for removal of an abandoned tank and the discovery of contaminated soil
- The discovery of a contamination plume (groundwater stream) which is subsequently traced back to the leaking tank through hydrogeologic testing

Leaks discovered through testing are usually the easiest to deal with because the extent of contamination tends to be small, the problem is contained at the site of the tank, and the volume of material to be removed is usually low.

Leaks found through excavation are often the messiest to deal with for the following reasons: First, an open excavation has been made into contaminated soil; contaminated soil has often already been removed from the excavation and stockpiled or shipped off-site; work is stopped to avoid risk to human health and safety; schedules are disrupted; and cleanup is done on an "as-needed" basis with no prior knowledge of the extent of costs. In short, the problem controls the situation, while people only react.

Leaks found through plume tracing are usually the most expensive to fix. Extensive subsoil and groundwater reclamation are generally required, at great cost. In addition, if the plume has been generated, or allowed to worsen through negligence of the tank owner, costs incurred by state and federal agencies in response to the plume are assessable against the tank owner. Fines are also imposed, and triple damages are assessed. From a technological standpoint, this type of leak may also present the most difficult cleanup problem due to the areal extent of the plume and the depth of soil contamination.

Protocol for Remediation

A classic approach to solving engineering problems also applies to the remediation of LUST. The steps involved are:

1. Define the problem.
2. Define all possible solutions.
3. Evaluate the possible solutions.
4. Select the best solution.
5. Implement the selected solution.

A LUST problem is, in fact, two problems: removal of the source tank, and remediation of the contamination. Both problems are addressed simultaneously with parallel efforts. Actual tank removal should not occur until the extent of the overall problem is known and an appropriate hazard evaluation completed.

Defining the Problem

From a technical standpoint, defining the problem is often the most difficult of the five steps. Before a remediation plan can be developed, it is necessary to know the areal extent of contamination, the nature of the contaminants, and the concentration of contaminants. The following five steps are used to generate this data.

1. Research the site history.
2. Excavate test pits and/or set drill holes and test wells in a predetermined pattern.
3. Test water and soil samples for pre-selected contaminants.
4. Plot isobars of contaminant concentration on a topographic map of the contaminated area.
5. Determine the level of concentration reduction required for each contaminant found.

Note that by determining the level of concentration reduction required for each contaminant found, one has defined the problem.

Researching the Site

The first step in defining the problem is to carefully and throroughly review the history of the site. Since a problem will already have been identified (a leaking underground storage tank is always a problem), and the source will have been found, the required site history information need only include the history of prior tank installations, prior tank removals, prior tank uses, and current tank uses. This history is important because the indicator pollutants to be sought during the plume location phase will depend upon the historical uses of the tank.

Test Pits

Once the indicator pollutants have been identified, it is time to begin a search for the plume of contamination. That can be done in several ways. The most conclusive method is to dig a series of test pits with a backhoe. As the backhoe excavates into the soil, the escaping air should be constantly monitored with a photoionization detector or flame ionization detector for the presence of organic and inorganic vapors. In addition, a combustible gas meter should be used to identify explosive gases and certain toxic compounds not ionized by the ionization detectors.

As the excavation proceeds, readings on the monitoring equipment will change. Readings should be made at least every five feet in depth, and the holes should be no more than 20′ apart near the suspected edge of the plume. Nearer the center of a large plume area, 100′ hole spacings, or greater, may be adequate and appropriate.

Whenever the monitoring equipment shows a reading greater than 10 ppm *above background*, a soil or water sample from that location and elevation should be sent to a laboratory for detailed analysis. Field monitoring equipment is usually not capable of either accurate analysis or compound differentiation.

Careful control of the samples taken is essential. Each sample should completely fill the sample jar, and the lid should be tightly closed. Some compounds require that samples be stabilized in the field if accurate laboratory results are to be obtained. A check with the laboratory technicians who will receive the samples is useful to ensure proper sample handling in the field.

Each sample also needs to be carefully marked or identified. Usually a test pit number and depth of sample location are adequate information for such labeling. If duplicate samples are taken, lower case letters may be used to differentiate the duplicates. The system itself is not too important, only that one exists and is carefully and consistently used.

A chain-of-custody record (See Figure 15.1) should be religiously maintained between the field and the laboratory. Custody of a sample means that the sample is *in the custodian's actual possession, in sight of the custodian, or secured in a locked, limited access area by the custodian*. The person in charge of field sampling is the first custodian because that person is the first to have physical possession. That custodian is responsible for marking the samples, stabilizing those requiring stabilization, securely storing the samples, and delivering them to the lab.

At the lab, the field custodian should receive a signed receipt for each sample. The laboratory custodian must then treat the samples in the same way as the field custodian, following the same rules of custodial care by physically maintaining control, or locking the samples in a secure, limited access area.

Laboratory Tests

Laboratory tests performed depend on the contaminants expected. Most commonly, at least some samples are subjected to EPA Method 624 testing. This test identifies the presence and concentration of 36 different volatile organics and chlorinated solvents. Where gasoline or petroleum spills are suspected, a test for total petroleum hydrocarbons (TPH) is useful. If the actual petroleum hydrocarbon spill is not clear, a hydrocarbon profile can be performed to identify the materials being found by the TPH test. If two or three scattered profiles indicate the same components, profiles need not be performed on other samples. Where specific compounds are known to have been stored in the LUST facilities, separate tests for those compounds are the best way to identify the plume.

Once the nature of the plume is known (i.e., the constituent contaminants have been identified and the concentrations in specific samples calculated), a relative soil conductivity curve can be developed. That curve will relate electrical conductivity of the soil to contaminant concentration. Thereafter, it is possible to use a field conductivity meter, rather than an ionization meter and laboratory results, to fully map the areal extent and depth of contamination by concentration levels, especially when the contaminants are in the groundwater.

Plot Isobars

When the laboratory and field data are all gathered, collected, and plotted on a topographic map of the area, a three-dimensional profile of contamination will result, on which isobars of equal concentration can be plotted. That three-dimensional model fully defines the extent of contamination.

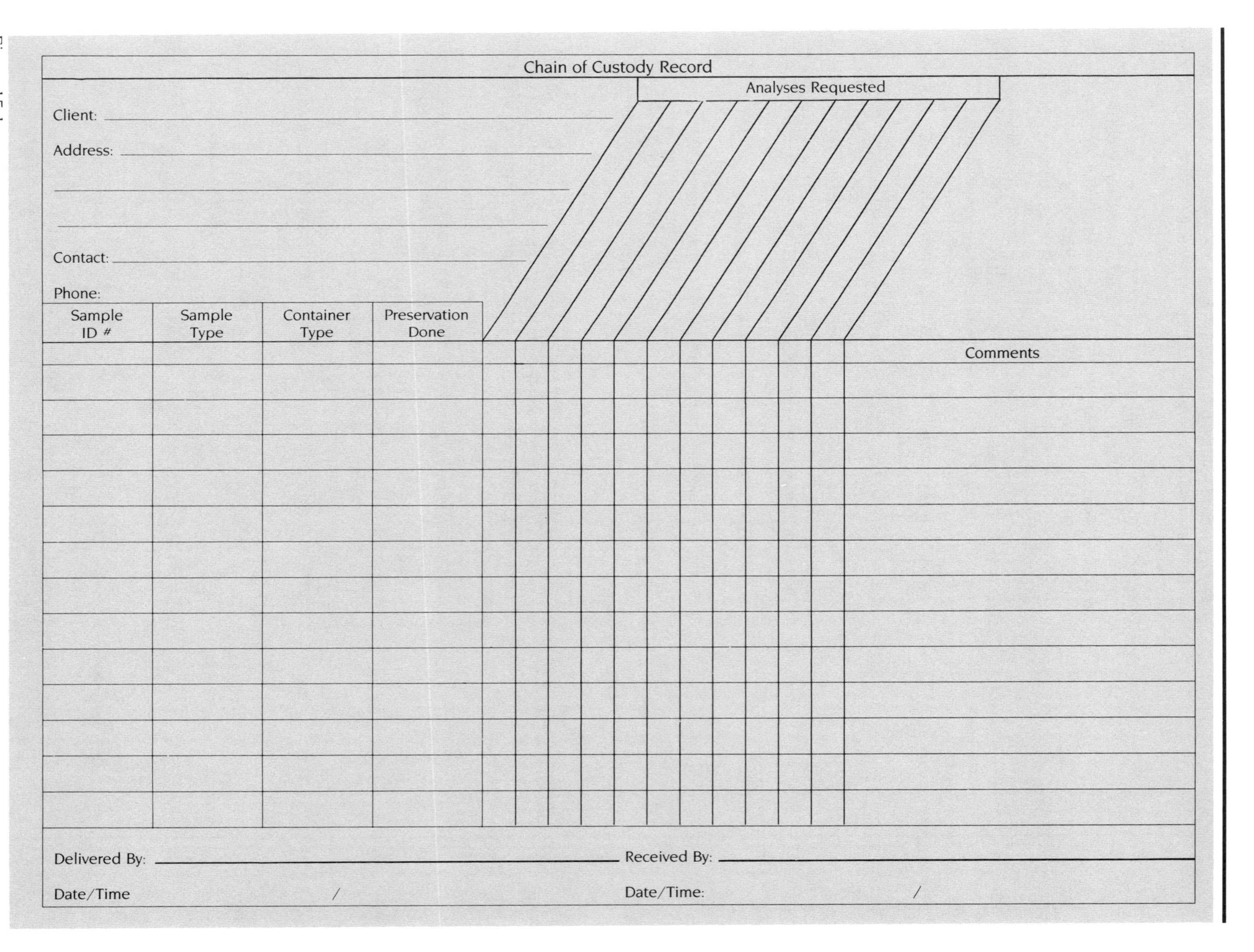

Chain of Custody Record

Client:

Address:

Contact:

Phone:

Sample ID #	Sample Type	Container Type	Preservation Done	Analyses Requested	Comments

Delivered By: Received By:

Date/Time / Date/Time: /

Figure 15.1

Determine Concentration Reduction Needs

Each contaminant being considered will affect people adversely at a different level of concentration. It is necessary to determine the Threshold Limit Value (TLV) of each contaminant. Standard chemical references provide this data. Some examples are listed below.

- CHRIS: Chemical Hazard Response Information System, developed by the U.S. Coast Guard. Access is through the National Response Center at (800) 424-8802. Copies are available for purchase through federal Bookstores and the Government Printing Office in Washington.
- *Documentation of the Threshold Limit Values* (TLV), fourth edition (1980), ACGIH Publications Office, 6500 Glenway Avenue, Building D-5, Cincinnati, Ohio 45221.
- Various on-line computer services that provide telephone/computer link access to a wide range of information on thousands of chemicals and compounds.

By evaluating the TLV against the known concentrations in the soil and water, the engineer can determine the contaminant reduction requirements. Those data define the problem.

Solutions

Cleanup of a LUST site is driven by two factors: the nature and extent of the contaminant plume, and the concentration reduction needs.

A small spill that has contaminated only a few cubic yards of material is usually best cleaned by excavation of the contaminated soil for disposal at a hazardous waste facility off-site. Where extensive groundwater contamination has occurred, the water is pumped, treated to remove pollutants, and returned to the aquifer. A variety of techniques are available for other kinds of spills. These include insitu encapsulation; biological treatment; excavation, treatment and re-emplacement; and small-scale pumping and cleansing. Each technique has its own application, and the engineer needs to consider each one on a case-by-case basis. In virtually all cases, however, removal of the source tank is mandated.

Tank Removal Protocol

The physical removal of the source tank is not to be taken lightly. The contamination around the tank poses serious health and safety risks for on-site personnel. Most hydrocarbons, if inhaled, are at least toxic, and many are carcinogenic. Most will explode at relatively low concentrations in air, and all are capable of displacing sufficient oxygen in the air to render it unbreathable. Therefore, removal of the tanks must be carried out in accordance with strict safety procedures.

The following steps should be taken to safely remove a LUST.

1. Open the tank vents and access ports.
2. Drain the tank.
3. Purge the tank with an inert gas.
4. If access to the inside can be safely gained, enter the tank and clean out all contaminants (using proper personal protective equipment for access to confined spaces).
5. Determine the appropriate personal protective equipment (PPE) required for on-site personnel.
6. Don PPE and carefully excavate the soil surrounding the tank.
7. Pull the tank and dispose.
8. Implement the predetermined site cleanup plan.
9. Install new tanks or close the excavation.

Cleaning the Tank

By opening the vents, draining the tank, purging the inside, and cleaning the tank, the possibility of an accidental explosion is dramatically reduced. It should be noted that sparks are the principal cause of explosions and fires during tank removal operations. Every time a backhoe digs into the soil, sparks are possible. If a tank is punctured during the process, an explosion becomes likely. If the tank is pre-cleaned and purged, however, even a puncture of the tank will not cause a problem.

Before entering a tank (which is, by definition, a confined space), it is important to follow several safety precautions. First, check the tank atmosphere for the presence of oxygen. If oxygen levels are not at least 19.5% inside the tank, other meter readings will have to be adjusted. Second, check for the presence of explosive or combustible gases and other toxic vapors. Unless the tank has previously been cleaned and has not subsequently been put back into service, the air inside will not be suitable for breathing, even with an air-purifying respirator. Normally, a supplied air respirator will be required for anyone entering the tank to clean it. Note that this is bulky equipment, but essential to safe operations. (See Chapter 5 for more information on these protective respiratory devices, and other personal protective equipment).

Lights lowered into the tank must be intrinsically safe to avoid sparks. In addition, cleaning equipment should be made of wood or fabric. Metal tools must be non-sparking brass or aluminum. Tank interiors should be washed, rather than chipped or hammered, to the maximum extent possible. Also, pumping is preferred to bucket removal for debris. The tank should be thoroughly cleaned and purged, in place, in the ground, before any excavation begins.

Tank Excavation

Normally, level C as defined by the EPA, PPE (personal protective equipment), is adequate for tank removal. Canisters are available to handle a wide range of organic and inorganic vapors, mists, and dusts at rather high concentrations. Occasionally, however, contaminants will be so concentrated, oxygen levels so depleted, or gases so toxic, that supplied air respirators will be required.

As the excavation begins, and throughout its duration, the excavated material must be constantly monitored for the presence and concentration of toxic gases, and for the oxygen content. Whenever the concentration of gases exceeds 80% of the safe working level (as determined by the respirator manufacturer), or when the oxygen level drops below 19.5%, operations must cease and the PPE reevaluated.

A site safety officer (SSO) trained in the use of PPE and the characteristics of the pollutants encountered must be on-site during all excavation activities. The SSO determines the level of PPE required and is responsible for safe work practices on the site.

Site Cleanups

After the leaking tank has been pulled and cleaned of hazardous pollutants on the outside, it can be treated as scrap metal and removed from the site. Then the site cleanup can begin. The site remediation plan is implemented while the excavation is still open and the SSO is still on the site. Proper PPE is worn, as determined by the SSO. The nature of the cleanup work, of course, depends upon the engineer who designs the cleanup plan. Following site cleanup, new tanks may be installed or the hole backfilled.

Summary

A great deal of preliminary work is required to effect safe LUST management. By carefully planning and executing the preliminary work, rather than blindly rushing in to pull a tank, the safety of the tank removal personnel can be ensured, and the environment protected from further harm.

APPENDIX

Table of Contents

EPA Hazardous Waste Codes

Code	Waste Description
Characteristic Hazardous Waste	
D001	Ignitable waste
D002	Corrosive waste
D003	Reactive waste
D004	Arsenic
D005	Barium
D006	Cadmium
D007	Chromium
D008	Lead
D009	Mercury
D010	Selenium
D011	Silver
D012	Endrin(1,2,3,4,10,10-hexachloro-1,7-epoxy-1,4,4a,5,6,7,8,8a-octahydro-1, 4-endo, endo-5,8-dimeth-ano-naphthalene)
D013	Lindane (1,2,3,4,5,6-hexa-chlorocyclohexane, gamma isomer)
D014	Methoxychlor (1,1,1-trichloro-2,2-bis [p-methoxyphenyl] ethane)
D015	Toxaphene ($C_{10}H_{10}Cl_8$, technical chlorinated camphene, 67-69 percent chlorine)
D016	2,4-D (2,4-dichlorophenoxyacetic acid)
D017	2,4,5-TP Silvex (2,4,5-trichlorophenoxypropionic acid)
Hazardous Waste from Nonspecific Sources	
F001	The following spent halogenated solvents used in degreasing: tetrachloroethylene, trichloroethylene, methylene chloride, 1,1,1-trichloroethane, carbon tetrachloride and chlorinated fluorocarbons and all spent solvent mixtures/blends used in degreasing containing, before use, a total of 10 percent or more (by volume) of one or more of the above halogenated solvents or those solvents listed in F002, F004, and F005; and still bottoms from the recovery of these spent solvents and spent solvent mixtures

EPA Hazardous Waste Codes	
Code	Waste Description
F002	The following spent halogenated solvents: tetrachloroethylene, methylene chloride, trichloroethylene, 1,1,1-trichloroethane, chlorobenzene, 1,1,2-trichloro-1,2,2-trifluoroethane, ortho-dichlorobenzene, trichlorofluoromethane, and 1,1,2, trichloroethane; all spent solvent mixtures/blends containing, before use, a total of 10 percent or more (by volume) of one or more of the above halogenated solvents or those solvents listed in F001, F004, and F005; and still bottoms from the recovery of these spent solvents and spent solvent mixtures
F003	The following spent nonhalogenated solvents: xylene, acetone, ethyl acetate, ethyl benzene, ethyl ether, methyl isobutyl ketone, n-butyl alcohol, cyclohexanone, and methanol; all spent solvent mixtures blends containing, before use, only the above spent nonhalogenated solvents; and all spent solvent mixtures/blends containing, before use, one or more of the above nonhalogenated solvents, and a total of 10 percent or more (by volume) of one or more of those solvents listed in F001, F002, F004, and F005; and still bottoms from the recovery of these spent solvents and spent solvent mixtures
F004	The following spent nonhalogenated solvents: cresols and cresylic acid, and nitrobenzene; and the still bottoms from the recovery of these solvents; all spent solvent mixtures/blends containing before use a total of 10% or more (by volume) of one or more of the above nonhalogenated solvents or those solvents listed in F001, F002, and F005; and still bottoms from the recovery of these spent solvents and spent solvent mixtures
F005	The following spent nonhalogenated solvents: toluene, methyl ethyl ketone, carbon disulfide, isobutanol, pyridine, benzene, 2-ethoxyethanol, and 2-nitropropane; all spent solvent mixtures/blends containing, before use, a total of 10 percent or more (by volume) of one or more of the above nonhalogenated solvents or those solvents listed in F001, F002, or F004; and still bottoms from the recovery of these spent solvents and spent solvent mixtures
F006	Wastewater treatment sludges from electroplating operations except from the following processes: (1) Sulfuric acid anodizing of aluminum; (2) tin plating on carbon steel; (3) zinc plating (segregated basis) on carbon steel; (4) aluminum or zinc-aluminum plating on carbon steel; (5) cleaning/stripping associated with tin, zinc, and aluminum plating on carbon steel; and (6) chemical etching and milling of aluminum

EPA Hazardous Waste Codes	
Code	Waste Description
F007	Spent cyanide plating bath solutions from electroplating operations
F008	Plating bath residues from the bottom of plating baths from electroplating operations where cyanides are used in the process
F009	Spent stripping and cleaning bath solutions from electroplating operations were cyanides are used in the process
F010	Quenching bath residues from oil baths from metal heat treating operations where cyanides are used in the process
F011	Spent cyanide solutions from slat bath pot cleaning from metal heat treating operations
F012	Quenching waste water treatment sludges from metal heat treating operations where cyanides are used in the process
F019	Wastewater treatment sludges from the chemical conversion coating of aluminum
F020	Wastes (except wastewater and spent carbon from hydrogen chloride purification) from the production or manufacturing use (as a reactant, chemical intermediate, or component in a formulating process) of tri- or tetrachlorophenol or of intermediates used to produce their pesticide derivatives. (This listing does not include wastes from the production of hexachlorophene from highly purified 2,4,5-trichlorophenol.)
F021	Wastes (except wastewater and spent carbon from hydrogen chloride purification) from the production or manufacturing use (as a reactant, chemical intermediate, or component in a formulating process) of pentachlorophenol, or of intermediates used to produce derivatives
F022	Wastes (except wastewater and spent carbon from hydrogen chloride purification) from the manufacturing use (as a reactant, chemical intermediate, or component in a formulating process) of tetra-, penta-or hexachlorobenzenes under alkaline conditions
F023	Wastes (except wastewater and spent carbon from hydrogen chloride purification) from the production of materials on equipment previously used for the production or manufacturing use (as a reactant, chemical intermediate, or component in a formulating process) of tri- and tetrachlorophenols. (This listing does not include wastes from equipment used only for the production or use of Hexachlorophene from highly purified 2,4,5-trichlorophenol.)

EPA Hazardous Waste Codes	
Code	Waste Description
F024	Wastes, including but not limited to, distillation residues, heavy ends, tars, and reactor clean-out wastes from the production of chlorinated aliphatic hydrocarbons, having a carbon content from one to five, utilizing free radical catalyzed processes. (This listing does not include light ends, spent filters and filter aids, spent dessicants, wastewater, wastewater treatment sludges, spent catalysts, and wastes listed in A 261.32.)
F026	Wastes (except wastewater and spent carbon from hydrogen chloride purification) from the production of materials on equipment previously used for the manufacturing use (as a reactant, chemical intermediate, or component in a formulating process) of tetra-, penta-, or hexachlorobenzene under alkaline conditions.
F027	Discarded unused formulations containing tri-, tetra-, or pentachlorophenol or discarded unused formulations containing compounds derived from these chlorophenols. (This listing does not include formulations containing hexachlorophene synthesized from prepurified 2,4,5-trichlorophenol as the sole component.)
F028	Residues resulting from the incineration or thermal treatment of soil contaminated with EPA hazardous waste nos. F020, F021, F022, F023, F026, and F027
Hazardous Waste from Specific Sources	
K001	Bottom sediment sludge from the treatment of wastewater from wood preserving processes that use creosote and/or pentachlorophenol
K002	Wastewater treatment sludge from the production of chrome yellow and orange pigments
K003	Wastewater treatment sludge from the production of molybdate orange pigments
K004	Wastewater treatment sludge from the production of zinc yellow pigments
K005	Wastewater treatment sludge from the production of chrome green pigments
K006	Wastewater treatment sludge from the production of chrome oxide green pigments (anhydrous and hydrated)
K007	Wastewater treatment sludge from the production of iron blue pigments
K008	Oven residue from the production of chrome oxide green pigments

EPA Hazardous Waste Codes	
Code	Waste Description
K009	Distillation bottoms from the production of acetaldehyde from ethylene
K010	Distillation side cuts from the production of acetaldehyde from ethylene
K011	Bottom stream from the wastewater stripper in the production of acrylonitrile
K013	Bottom stream from the acetonitrile column in the production of acrylonltrile
K014	Bottoms from the acetonitrile purification column in the production of acrylonitrile
K015	Still bottoms from the distillation of benzyl chloride
K016	Heavy ends or distillation residues from the production of carbon tetrachloride
K017	Heavy ends (still bottoms) from the purification column in the production of epichlorohydrin
K018	Heavy ends from the fractionation column in ethyl chloride production
K019	Heavy ends from the distillation of ethylene dichloride in ethylene dichloride production
K020	Heavy ends from the distillation of vinyl chloride in vinyl chloride monomer production
K021	Aqueous spent antimony catalyst waste from fluoromethanes production
K022	Distillation bottom tars from the production of phenol/ acetone from cumene
K023	Distillation light ends from the production of phthalic anhydride from naphthalene
K024	Distillation bottoms from the production of phthalic anhydride from naphthalene
K025	Distillation bottoms from the production of nitrobenzene by the nitration of benzene
K026	Stripping still tails from the production of methyl ethyl pyridines
K027	Centrifuge and distillation residues from toluene diisocyanate production
K028	Spent catalyst from the hydrochlorinator reactor in the production of 1,1,1-trichloroethane
K029	Waste from the product steam stripper in the production of 1,1,1-trichloroethane
K030	Column bottoms or heavy ends from the combined production of trichloroethylene and perchloroethylene
K031	By-product salts generated in the production of MSMA and cacodylic acid

EPA Hazardous Waste Codes	
Code	Waste Description
K032	Wastewater treatment sludge from the production of chlordane
K033	Wastewater and scrub water from the chlorination of cyclopentadiene in the production of chlordane
K034	Filter solids from the filtration of hexachlorocyclopentadiene in the production of chlordane
K035	Wastewater treatment sludges generated in the production of creosote
KO36	Still bottoms from toluene reclamation distillation in the production of disulfoton
K037	Wastewater treatment sludges from the production of disulfoton
K038	Wastewater from the washing and stripping of phorate production
K039	Filter cake from the filtration of diethylphosphorodithioic acid in the production of phorate
K040	Wastewater treatment sludge from the production of phorate
K041	Wastewater treatment sludge from the production of toxaphene
K042	Heavy ends or distillation residues from the distillation of tetrachlorobenzene in the production of 2,4,5-T
K043	2,6-Dichlorophenol waste from the production of 2,4-D
K044	Wastewater treatment sludges from the manufacturing and processing of explosives
K045	Spent carbon from the treatment of wastewater containing explosives
K046	Wastewater treatment sludges from the manufacturing, formulation and loading of lead-based initiating compounds
K047	Pink/red water from TNT operations
K048	Dissolved air flotation (DAF) float from the petroleum refining industry
K049	Slop oil emulsion solids from the petroleum refining industry
K050	Heat exchanger bundle cleaning sludge from the petroleum refining industry
K051	API separator sludge from the petroleum refining industry
K052	Tank bottoms (leaded) from the petroleum refining industry

EPA Hazardous Waste Codes

Code	Waste Description
K060	Ammonia still lime sludge from coking operations
K061	Emission control dust/sludge from the primary production of steel in electric furnaces.
K062	Spent pickle liquor from steel finishing operations of plants that produce iron or steel
K069	Emission control dust/sludge from secondary lead smelting
K071	Brine purification muds from the mercury cell process in chlorine production, where separately prepurified brine is not used
K073	Chlorinated hydrocarbon waste from the purification step of the diaphragm cell process using graphite anodes in chlorine production
K083	Distillation bottoms from aniline production
K084	Wastewater treatment sludges generated during the production of veterinary pharmaceuticals from arsenic or organo-arsenic compounds
K085	Distillation or fractionation column bottoms from the production of chlorobenzenes
K086	Solvent washes and sludges, caustic washes and sludges, or water washes and sludges from cleaning tubs and equipment used in the formulation of ink from pigments, driers, soaps, and stabilizers containing chromium and lead
K087	Decanter tank tar sludge from coking operations
K093	Distillation light ends from the production of phthalic anhydride from ortho-xylene
K094	Distillation bottoms from the production of phthalic anhydride from ortho-xylene
K095	Distillation bottoms from the production of 1,1,1-trichloroethane
K096	Heavy ends from the heavy ends column from the production of 1,1,1-trichloroethane
K097	Vacuum stripper discharge from the chlordane chlorinator in the production of chlordane
K098	Untreated process wastewater from the production of toxaphene
K099	Untreated wastewater from the production of 2,4-D
K100	Waste leaching solution from acid leaching of emission control dust/sludge from secondary lead smelting
K101	Distillation tar residues from the distillation of aniline-based compounds in the production of veterinary pharmaceuticals from arsenic or organo-arsenic compounds

EPA Hazardous Waste Codes	
Code	**Waste Description**
K102	Residue from the use of activated carbon for decolorization in the production of veterinary pharmaceuticals from arsenic or organo-arsenic compounds
K103	Process residues from aniline extraction from the production of aniline
K104	Combined wastewater streams generated from nitrobenzene/aniline production
K105	Separated aqueous stream from the reactor product washing step in the production of chlorobenzenes
K106	Wastewater treatment sludge from the mercury cell process in chlorine production
K123	Process wastewater (including supernates, filtrates, and wash waters) from the production of Ethylenebisdithiocarbamic Acids and its salt. Hazardous Code T
K124	Reactor vent scrubber water from the production of Ethylenebisdithiocarbamic and its salts. Hazardous Code T
K125	Filtration, evaporation, and centrifugation of solids from the production of Ethylenebisdithiocarbamic Acids and its salts. Hazardous Code T and C
K126	Baghouse dust and floor sweepings in milling and packaging operations from production or formulation of Ethylenebisdithiocarbamic Acids and its salts. Hazardous Code T
K111	Product washwaters from the production of dinitrotoluene via nitration of toluene
K112	Reaction byproduct water from the drying column in the production of toluenediamine via hydrogenation of dinitrotoluene
K113	Condensed liquid light ends from purification of toluenediamine in production of toluenediamine via hydrogenation of dinitrotoluene.
K114	Vicinals from the purification of toluenediamine in production of toluenediamine via hydrogenation of dinitrotoluene
K115	Heavy ends from purification of toluenediamine in the production of toluenediamine via hydrogenation of dinitrotoluene
K116	Organic condensate from the solvent recovery column in the production of toluene diisocyanate via phosgenation of toluenediamine
K117	Wastewater from the reactor vent gas scrubber in the production of ethylene dibromide via bromination of ethene

EPA Hazardous Waste Codes	
Code	**Waste Description**
K118	Spent adsorbent solids from purification of ethylene dibromide in the production of ethylene dibromide via bromination of ethene
K136	Still bottoms from the purification of ethylene dibromide in the production of ethylene dibromide via bromination of ethene

Discarded Commercial Chemical Products, Off-Specification Species, Container Residuals, and Spill Residues Thereof—Acute Hazardous Waste
(*An alphabetized listing can be found at* 40 CFR 261.33, *July* 1, 1986.)

Code	Waste Description
P001	Warfarin, when present at concentrations greater than or equal to 0.3%
P001	3-(alpha-Acetonyl-benzyl)-4-hydroxycoumarin and salts, when present at concentrations greater than 0.3%
P002	Acetamide, N-(aminothioxomethyl)
P002	1-Acetyl-2-thiourea
P003	2-Propenal
P003	Acrolein
P004	1,2,3,4,10,10-Hexachloro-1,4,4a,5,8,8a-hexahydro-1,4:5,8-endo, exo-dimethanonapthalene
P004	Aldrin
P005	2-Propen-1-ol
P005	Allyl alcohol
P006	Aluminum phosphide (r,t)
P007	3(2H)-Isoxazolone,5-(aminomethyl)-
P007	5-(Aminomethyl)-3-isoxazolol
P008	4-a-Aminopyridine
P008	4-Pyridinamine
P008	4-Aminopyridine
P009	Phenol,2,4,6-trinitro-,ammonium salt (r)
P009	Ammonium picrate (r)
P010	Arsenic acid (t)
P011	Arsenic pentoxide (t)
P011	Arsenic (V) oxide (t)
P012	Arsenic (III) oxide (t)
P012	Arsenic trioxide (t)
P013	Barium cyanide
P014	Thiophenol
P014	Benzenethiol
P015	Beryllium dust (t)

EPA Hazardous Waste Codes	
Code	**Waste Description**
P016	Methane,oxybis(chlor-
P016	Bis(chloromethyl) ether
P017	2-Propanone,1-bromo-(t)
P017	Bromoacetone (t)
P018	Strychnidinone,2,3-dimethoxy-
P018	Brucine
P020	Dinoseb
P020	Phenol,2,4dinitro-6-(1-methylpropyl)
P021	Calcium cyanide
P022	Carbon bisulfide (t)
P022	Carbon disulfide (t)
P023	Acetaldehyde, chloro-
P023	Chloroacetaldehyde
P024	Benzenamine, 4-chloro
P024	p-Chloroaniline
P026	Thiourea, (2-chlorophenyl)-
P026	1-(o-Chlorophenyl)thiourea
P027	Propanenitrile,3-chloro-
P027	3-Chloropropionitrile
P028	Benzene, (chloromethyl)-
P028	Benzyl chloride
P029	Copper cyanides
P030	Cyanides (soluble cyanide salts), not elsewhere specified (t)
P031	Cyanogen
P033	Cyanogen chloride
P033	Chlorine cyanide
P034	4,6-Dinitro-0-cyclohexylphenol (t)
P034	Phenol,2-cyclohexyl-4,6-dinitro-(t)
P036	Dichlorophenylarsine
P036	Phenyl dichloroarsine
P037	Dieldrin
P037	1,2,3,4,10,10-Hexachloro-6,7-expoxy-1,4,4a,5,6,7,8,8a-octahydro-endo, exo-1,4:5,8-dimethanonaphthalene
P038	Diethylarsine (t)
P038	Arsine, diethyl- (t)
P039	0,0-Diethyl S-[2-(ethylthio)ethyl phosphorodithioate (t)
P039	Disulfoton (t)

EPA Hazardous Waste Codes	
Code	Waste Description
P040	0,0-Diethyl 0-pyrazinyl phosphorothioate
P040	Phosphorothioic acid, 0,0-diethyl 0-pyrazinyl ester
P041	Diethyl-p-nitrophenyl phosphate
P041	Phosphoric acid, diethyl p-nitrophenyl ester
P042	Epinephrine
P042	1,2-Benzenediol, 4-[1-hydroxy-2-(methylamino)ethyl]-
P043	Diisopropyl fluorophosphate
P043	Fluoridic acid, bis(1-methylethyl) ester
P043	Phosphorofluoridic acid, bis(1-methylethyl) ester
P044	Dimethoate (t)
P044	Phosphorodithioic acid, 0,0-dimethyl S-[2-(methylamino)-2-oxoethyl]ester (t)
P045	3,3-Dimethyl-1-(methylthio)-2-butanone, 0-[(methylamino)carbonyl]oxime
P045	Thiofanox
P046	alpha,alpha-Dimethylphenethylamine (t)
P046	Ethanamine,1,1-dimethyl-2-phenyl-(t)
P047	4,6-Dinitro-o-cresol and salts
P047	Phenol,2'4-dinetro-6-methyl-, and salts
P048	2,4-Dinitrophenol
P048	Phenol,2,4-dinitro-
P049	2,4-Dithiobiuret
P049	Thiomidodicarbonic diamide
P050	Endosulfan
P050	5-Norbornene-2,3-dimethanol,1,4,5,6,7,7-hexachloro,cyclic sulfite
P051	1,2,3,4,10,10-Hexachloro-6,7expoxy-1,4,4a,5,6,7,8,8a-oxtahydro-endo,endo-1,4:5,8-dimethanon-aphthalene
P051	Endrin
P054	Ethylenimine
P054	Aziridine
P056	Fluorine
P057	Fluoroacetamide
P057	Acetamide,2-fluor-
P058	Fluoroacetic acid, sodium salt
P058	Acetic acid, fluoro-, sodium salt
P059	Heptachlor
P059	4,7-Methano-1H-indene,1,4,5,6,7,8,8-heptachloro-3a,4,7,7a-tetrahydro-

EPA Hazardous Waste Codes	
Code	**Waste Description**
P060	Hexachlorohexahydro-endo,endo-dimethanonapthalene
P060	1,2,3,4,10,10-Hexachloro-1,4,4a,5,8,8a-hexahydro-1,4:5,8-endo, endo-dimethanonaphthalene
P062	Hexaethyl tetraphosphate
P062	Tetraphosphoric acid, hexaethyl ester
P063	Hydrocyanic acid
P063	Hydrogen cyanide
P064	Methyl isocyanate
P064	Isocyanic acid, methyl ester
P065	Fulminic acid, mercury(II) salt (r,t)
P065	Mercury fulminate (r,t)
P066	Methomyl
P066	Acetimidic acid, N-[(methylcarbamoyl)oxy]thio-, methyl ester
P067	2-Methylaziridine
P067	1,2-Propylenimine
P068	Hydrazine,methyl-
P068	Methyl hydrazine
P069	2-Methyllactonitrile
P069	Propanenitrile,2-hydroxy-2-methyl
P070	Propanal, 2-methyl-2-(methylthlo)-, 0[(methyl-amino)carbonyl]oxime
P070	Aldicarb
P071	0,0-Dimethyl 0-p-nitrophenyl phosphorothioate
P071	Methyl parathion
P072	alpha-Naphthylthiourea
P072	Thiourea, 1-naphthalenyl
P073	Nickel tetracarbonyl
P073	Nickel carbonyl
P074	Nickel(II)cyanide
P074	Nickel cyanide
P075	Nicotine and salts (t)
P075	Pyridine, (S)-3-(1-methyl-2-pyrrolidinyl)-, and salts
P076	Nitrogen (II) oxide (t)
P076	Nitric oxide (t)
P077	p-Nitroaniline (t)
P077	Benzenamine, 4-nitro-
P078	Nitrogen (IV) oxide

EPA Hazardous Waste Codes	
Code	Waste Description
P078	Nitrogen dioxide
P081	Nitroglycerine (r,t)
P081	1,2,3-Propanetriol,trinitrate-(r)
P082	Dimethylnitrosamine
P082	N-Nitrosodimethylamine
P084	Ethenamine,N-methyl-N-nitroso-
P084	N-Nitrosomethylvinylamine
P085	Diphosphoramide,octamethyl
P085	Octamethylpyrophosphoramide
P087	Osmium tetroxide
P087	Osmium oxide
P088	Endothall
P088	7-Oxabicyclo[2.2.1]heptane-2,3-dicarboxylic acid
P089	Parathion (t)
P089	Phosphorothioic acid,0,0-diethyl 0-(p-nitrophenyl) ester (t)
P092	Mercury,(acetato-0)phenyl-
P092	Phenylmercuric acetate
P093	N-Phenylthiourea
P093	Thiourea, phenyl-
P094	Phosphorothioic acid, 0,0-diethyl S-(ethylthio)methyl ester (t)
P094	Phorate (t)
P095	Phosgene (t)
P095	Carbonyl chloride
P096	Hydrogen phosphide
P096	Phosphine
P097	Famphur
P097	Phosphorothioic acid, 0,0-dimethyl 0-[p-((dimethylamino)-sulfonyl)phenyl]ester
P098	Potassium cyanide
P099	Potassium silver cyanide
P101	Ethyl cyanide
P101	Propanenitrile
P102	Propargyl alcohol
P102	2-Propyn-1-ol
P103	Selenourea
P103	Carbamimidoselenoic acid
P104	Silver cyanide

EPA Hazardous Waste Codes	
Code	**Waste Description**
P105	Sodium azide
P106	Sodium cyanide
P107	Strontium sulfide (t)
P108	Strychnidin-10-one, and salts (t)
P108	Strychnine and salts (t)
P109	Dithiopyrophosphoric acld, tetraethyl ester
P109	Tetraethyldithiopyrophosphate
P110	Plumbane,tetraethyl-
P110	Tetraethyl lead
P111	Tetraethylpyrophosphate
P111	Pyrophosphoric acid, tetraethyl ester
P112	Methane,tetranitro-(r)
P112	Tetranitromethane (r)
P113	Thallium(III) oxide
P113	Thallic oxide
P114	Thallium (I) selenide
P115	Sulfuric acid, thallium (I) salt
P115	Sulfuric acid, thallium(I) salt
P115	Thallium(I)sulfate
P116	Hydrazinecarbothioamide
P116	Thiosemicarbazide
P118	Methanethiol,trichloro-
P118	Trichloromethanethiol
P119	Vanadic acid, ammonium salt
P119	Ammonium vanadate
P120	Vanadium pentoxide
P120	Vanadium(V) oxide
P121	Zinc cyanide
P122	Zinc phosphide (r,t)
P122	Zinc phosphide, when present at concentrations greater than 10%
P123	Toxaphene
P123	Camphene, octachloro-

Discarded Commercial Chemical Products, Off-Specification Species, Container Residues, and Spill Residues Thereof—Toxic Waste
(*An alphabetized listing can be found at* 40 CFR 261.33, *July* 1, 1986.)

U001	Ethanal (i)
U001	Acetaldehyde (i)

EPA Hazardous Waste Codes	
Code	Waste Description
U002	2-Propanone (i)
U002	Acetone (i)
U003	Ethanenitrile (i,t)
U003	Acetonitrile (i,t)
U004	Ethanone,1-phenyl-
U004	Acetophenone
U005	2-Acetylaminofluorene
U005	Acetamide, N-9H-fluoren-2-yl-
U006	Ethanoyl chloride (c,r,t)
U006	Acetyl chloride (c,r,t)
U007	2-Propenamide
U007	Acrylamide
U008	2-Propenoic acid (i)
U008	Acrylic acid (i)
U009	2-Propenenitrile
U009	Acrylonitrile
U010	Mitomycin C
U010	Azirino(2'3':3,4)pyrrolo(1,2-a) indole-4,7-dione, 6-amino-8-[((aminocarbonyl) oxy)methyl]-1,1a,2,8,8a,8b-hexahydro-8a-methoxy-5-methyl-,
U011	1H-1,2,4-Triazol-3-amine
U011	Amitrole
U012	Benzenamine (i,t)
U012	Aniline (i,t)
U014	Auramine
U014	Benzenamine, 4,4'-carbonimidoylbis(N,N-dimethyl-
U015	L-Serine, diazoacetate (ester)
U015	Azaserine
U016	Benz[c]acridine
U016	3,4-Benzacridine
U017	Benzal chloride
U017	Benezene, (dichloromethyl)-
U018	Benz[a]anthracene
U018	1,2-Benzanthracene
U019	Benzene (i,t)
U020	Benzenesulfonyl chloride (c,r)
U020	Benzenesulfonic acid chloride (c,r)
U021	Benzidine

EPA Hazardous Waste Codes	
Code	Waste Description
U021	(1,1'-Biphenyl)-4,4'-diamine
U022	Benzo[a]pyrene
U022	3,4-Benzopyrene
U023	Benzotrichloride (c,r,t)
U023	Benzene, (trichloromethyl)(c,r,t)
U024	Bis(2-chloroethoxy) methane
U024	Ethane,1,1'-[methylenebis(oxy)]bis[2-chloro-
U025	Dichloroethyl ether
U025	Ethane,1,1'-oxybis[2-chloro-
U026	2-Naphthylamine,N,N-bis(2-chloromethyl)-
U026	Chlornaphazine
U027	Propane,2,2'-oxybis[2-chloro-
U027	Bis(2-chloroisopropyl) ether
U028	Bis(2-ethylhexyl) phthalate
U028	1,2-Benzenedicarboxylic acid, [bis(2-ethylhexyl)]ester
U029	Methane, bromo-
U029	Methyl bromide
U030	4-Bromophenyl phenyl ether
U030	Benzene, 1-bromo-4-phenoxy-
U031	1-Butanol (i)
U031	N-Butyl alchohol (i)
U032	Calcium chromate
U032	Chromic acid, calcium salt
U033	Carbonyl fluoride (r,t)
U033	Carbon oxyfluoride (r,t)
U034	Chloral
U034	Acetaldehyde, trichloro-
U035	Butanoic acid, 4-[bis(2-chloroethyl)amino]benzene-
U035	Chlorambucil
U036	4,7-Methanoindan, 1,2,4,5,6,7,8,8-octa-chloro-3a,4,7,7a-tetrahydro-
U036	Chlordane, technical
U037	Chlorobenzene
U037	Benzene, chloro-
U038	Ethyl 4,4,'-dichlorobenzilate
U038	Benzeneacetic acid, 4-chloro-alpha-4-chloro-phenyl)-alpha-hydroxy, ethyl ester
U039	Phenol,4-chloro-3-methyl-

EPA Hazardous Waste Codes	
Code	Waste Description
U039	4-Chloro-m-cresol
U041	Oxirane,2-(chloromethyl)-
U041	1-Chloro-2,3-expoxypropane
U042	Ethene,2-chloroethoxy-
U042	2-Chloroethyl vinyl ether
U043	Ethene,chloro
U043	Vinyl chloride
U044	Methane, trichloro-
U044	Chloroform
U045	Methane, chloro-(i,t)
U045	Methyl chloride (i,t)
U046	Methane, chloromethoxy-
U046	Chloromethyl methyl ether
U047	Naphthalene, 2-chloro-
U047	beta-Chloronaphthalene
U048	Phenol,2-chloro-
U048	o-Chlorophenol
U049	4-Chloro-o-toluidine,hydrochloride
U049	Benzenamine, 4-chloro-2-methyl-
U050	1,2-Benzphenanthrene
U050	Chrysene
U051	Creosote
U052	Cresylic acid
U052	Cresols
U053	2-Butenal
U053	Crotonaldehyde
U055	Cumene (i)
U055	Benzene, (1-methylethyl)-(i)
U056	Cyclohexane (i)
U056	Benzene, hexahydro- (i)
U057	Cyclohexanone (i)
U058	2H-1,3,2-Oxazaphosphorine, 2-[bis(2-chloroethyl)amino]-tetrahydro-2 oxide
U058	Cyclophosphamide
U059	5,12-Naphthacenedione, (8S-cis)-8-acetyl[3-amino-2,3,6-trideoxy-alpha-L-lyxohexopyranosyl)oxyl] 7,8.9.10-tetrahydro-6,8,11-trihydroxy-1-methoxy-
U059	Daunomycin
U060	Dichloro diphenyl dichloroethane

EPA Hazardous Waste Codes	
Code	Waste Description
U060	DDD
U061	DDT
U061	Dichloro diphenyl trichloroethane
U062	Diallate
U062	S-(2,3-Dichloroallyl) diisopropylthiocarbamate
U063	Dibenz[a,h]anthracene
U063	1,2:5,6-Dibenzanthracene
U064	Dibenz[a,i]pyrene
U064	1,2:7,8-Dibenzopyrene
U066	Propane,1-2-dibromo-3-chloro-
U066	1,2-Dibromo-3-chloropropane
U067	Ethylene dibromide
U067	Ethane, 1,2-dibromo-
U068	Methane, dibromo-
U068	Methylene bromide
U069	Dibutyl phthalate
U069	1,2-Benzenedicarboxylic acid, dibutyl ester
U070	o-Dichlorobenzene
U070	Benzene, 1,2-dichloro-
U071	m-Dichlorobenzene
U071	Benzene, 1,3-dichloro-
U072	p-Dichlorobenzene
U072	Benzene, 1,4-dichloro
U073	(1,1'-Biphenyl)-4,4'-diamine,3,3'-dichloro
U073	3,3'-Dichlorobenzidine
U074	2-Butene,1,4-dichloro-(i,t)
U074	1,4-Dichloro-2-butene (i,t)
U075	Methane, dichlorodifluoro-
U075	Dichlorodifluoromethane
U076	Ethylidene dichloride
U076	Ethane,1,1-dichloro-
U077	Ethylene dichloride
U077	Ethane,1,2-dichloro-
U078	Ethene,1-1-dichloro-
U078	1,1-Dichloroethylene
U079	Ethene, trans-1,2-dichloro-
U079	1,2-Dichloroethylene
U080	Methane, dichloro-

EPA Hazardous Waste Codes	
Code	Waste Description
U080	Methylene chloride
U081	Phenol,2,4-dichloro-
U081	2,4-Dichlorophenol
U082	Phenol,2,6-dichloro-
U082	2,6-Dichlorophenol
U083	Propylene dichloride
U083	1,2-Dichloropropane
U084	Propene,1,3-dichloro-
U084	1,3-Dichloropropene
U085	2,2'-Bioxirane (i,t)
U085	1,2:3,4-Diepoxybutane (i,t)
U086	Hydrazine, 1,2-diethyl-
U086	N,N-Diethylhydrazine
U087	Phosphorodithioic acid,0,0-diethyl-, S-methyl-ester
U087	0,0-Diethyl-S-methyl-dithiophosphate
U088	Diethyl phthalate
U088	1,2-Benzenedicarboxylic acid, diethyl ester
U089	4,4', Stilbenediol,alpha,alpha'-diethyl
U089	Diethylstilbestrol
U090	Dihydrosafrole
U090	Benzene,1,2-methylenedioxy-4-propyl-
U091	(1,1'-Biphenyl)-4,4'-diamine,3,3'-dimethoxy-
U091	3,3'-Dimethoxybenzidine
U092	Methanamine, N-methyl-(i)
U092	Dimethylamine (i)
U093	Dimethylaminoazobenzene
U093	Benzenamine,N,N-dimethyl-4-phenylazo-
U094	7,12-Dimethylbenz[a]anthracene
U094	1,2-Benzanthracene,7,12-dimethyl-
U095	(1,1'-Biphenyl)-4,4'-diamine,3,3'-dimethyl-
U095	3,3'-Dimethylbenzidine
U096	Hydroperoxide, 1-methyl-phenylethyl-(r)
U096	alpha,alpha-Dimethylbenzylhydroperoxide (r)
U097	Carbamoyl chloride,dimethyl-
U097	Dimethylcarbamoyl chloride
U098	Hydrazine,1,1-dimethyl-
U098	1,1-Dimethylhydrazine
U099	Hydrazine, 1,2-dimethyl-

EPA Hazardous Waste Codes	
Code	Waste Description
U099	1,2-Dimethylhydrazine
U101	Phenol,2,4-dimethyl-
U101	2,4-Dimethylphenol
U102	Dimethyl phthalate
U102	1-2-Benzenedicarboxylic acid, dimethyl ester
U103	Sulfuric acid, dimethyl ester
U103	Dimethyl sulfate
U105	2,4-Dinitrotoluene
U105	Benzene, 1-methyl-2,4-dinitro-
U106	2,6-Dinitrotoluene
U106	Benzene, 1-methyl-2,6-dinitro
U107	Di-n-octyl phthalate
U107	1-2-Benzenedicarboxylic acid, di-n-octyl ester
U108	1,4-Diethylene dioxide
U108	1,4-Dioxane
U109	Hydrazine, 1,2-diphenyl-
U109	1,2-Diphenylhydrazine
U110	1-Propanamine,N-propyl-(i)
U110	Dipropylamine (i)
U111	N-Nitroso-N-propylamine
U111	Di-N-propylnitrosamine
U112	Ethyl acetate (i)
U112	Acetic acid, ethyl ester (i)
U113	2-Propenoic acid, ethyl ester (i)
U113	Ethyl acrylate (i)
U114	Ethylenebis(dithiocarbamic acid), salts and esters
U114	1,2-Ethanediylbiscarbamodithoic acid
U115	Oxirane (i,t)
U115	Ethylene oxide (i,t)
U116	Ethylene thiourea
U116	2-Imidazolidinethione
U117	Ethyl ether (1)
U117	Ethane,1,1'-oxybis- (i)
U118	2-Propenoic acid, 2-methyl-, ethyl ester
U118	Ethyl methacrylate
U119	Ethyl methanesulfonate
U119	Methanesulfonic acid, ethyl ester
U120	Fluoranthene

EPA Hazardous Waste Codes	
Code	Waste Description
U120	Benzo[j,k]fluorene
U121	Trichloromonofluoromethane
U121	Methane, trichlorofluoro-
U122	Formaldehyde
U122	Methylene oxide
U123	Formic acid (c,t)
U123	Methanoic acid (c,t)
U124	Furan (i)
U124	Furfuran (i)
U125	Furfural(i)
U125	2-Furancarboxaldehyde (i)
U126	1-Propanol,2,3-epoxy-
U126	Glycidylaldehyde
U127	Hexachlorobenzene
U127	Benzene, hexachloro-
U128	Hexachlorobutadene
U128	1,3-Butadiene,1,1,2,3,4,4-hexachloro-
U129	Hexachlorocyclohexane (gamma isomer)
U129	Lindane
U130	Hexachlorocyclopentadene
U130	1,3-Cyclopentadiene,1,2,3,4,5,5-hexa- chloro-
U131	Hexachloroethane
U131	Ethane,1,1,1,2,2,2-hexachloro-
U132	Hexachlorophene
U132	2,2-Methylenebis(3,4,6-trichlorophenol)
U133	Hydrazine (r,t)
U133	Diamine (r,t)
U134	Hydrogen fluoride (c,t)
U134	Hydrofluoric acid(c,t)
U135	Sulfur hydride
U135	Hydrogen sulfide
U136	Hydroxydimethylarsine oxide
U136	Cacodylic acid
U137	1,10-(1,2-Phenylene)pyrene
U137	Ideno[1,2,3-cd]pyrene
U138	Methane, iodo-
U138	Methyl iodide
U139	Ferric dextran

EPA Hazardous Waste Codes	
Code	Waste Description
U139	Iron dextran
U140	1-Propanol,2-methyl- (i,t)
U140	Isobutyl alcohol (i,t)
U141	Isosafrole
U141	Benzene, 1,2-methylenedioxy-4-propenyl-
U142	Kepone
U142	Decachlorooctahydro-1,3,4-metheno-2H-cyclobuta[c,d]-pentalen-2-one
U143	Lasiocarpine
U144	Lead acetate
U144	Acetic acid, lead salt
U145	Phosphoric acid, lead salt
U145	Lead phosphate
U146	Lead subacetate
U147	2,5-Furandione
U147	Maleic anhydride
U148	Maleic hydrazide
U148	1,2-Dihydro-3,6-pyradizinedione
U149	Propanedinitrile
U149	Malononitrile
U150	Melphalan
U150	Alanine, 3[p-bis(2-chloroethyl)amino] phenyl-,L-
U151	Mercury
U152	Propenenitrile,2-methyl- (i,t)
U152	Methacrylonitrile (i,t)
U153	Thiomethanol (i,t)
U153	Methanethiol (i,t)
U154	Methanol (i)
U154	Methyl alcohol (i)
U155	Pyridine, 2-[(2-dimethylamino)ethyl]-2-phenylamino-
U155	Methapyrilene
U156	Methyl chlorocarbonate (i,t)
U156	Carbonochloridic acid, methyl ester (i,t)
U157	3-Methylcholanthrene
U157	Benz[j]aceanthrylene, 1,2-dihydro-3-methyl-
U158	4,4'-Methylenebis(2-chloroaniline)
U158	Benzenamine,4,4'-methylenebis(2-chloro-
U159	Methyl ethyl ketone (i,t)

EPA Hazardous Waste Codes	
Code	Waste Description
U159	2-Butanone (i,t)
U160	Methyl ethyl ketone peroxide (r,t)
U160	2-Butanone peroxide (r,t)
U161	4-Methyl-2-pentanone (i)
U161	Methyl isobutyl ketone (i)
U162	2-Propenoic acid, 2-methyl-, methyl ester (i,t)
U162	Methyl methacrylate (i,t)
U163	Guanidine, N-nitroso-N-methyl-N'-nitro-
U163	N-methyl-N'-nitro-N-nitrosoguanidine
U164	4(1H)-Pyrimidinone, 2,3,-dihydro-6-methyl-2-thioxo-
U164	Methylthiouracil
U165	Naphthalene
U166	1,4,Naphthoquinone
U166	1,4,Naphthalenedione
U167	1-Naphthylamine
U167	alpha-Naphthylamine
U168	2-Naphthylamine
U168	beta-Naphthylamine
U169	Nitrobenzene (i,t)
U169	Benzene, nitro- (i,t)
U170	Phenol,4-nitro-
U170	P-Nitrophenol
U171	Propane,2-nitro-(i,t)
U171	2-Nitropropane (i,t)
U172	N-Nitrosodi-N-butylamine
U172	1-Butanamine, N-butyl-N-nitroso-
U173	Ethanol,2,2-(nitrosoimino)bis-
U173	N-Nitrosodiethanolamine
U174	N-Nitrosodiethylamine
U174	Ethanamine, N-ethyl-N-nitroso-
U176	N-Nitroso-N-ethylurea
U176	Carbamide,N-ethyl-N-nitroso-
U177	N-Nitroso-N-methylurea
U177	Carbamide,N-methyl-N-nitroso-
U178	N-Nitroso-N-methylurethane
U178	Carbamic acid, methylnitroso-, ethyl ester
U179	N-Nitrosopiperidine
U179	Pyridine,hexahydro-N-nitroso-

EPA Hazardous Waste Codes	
Code	Waste Description
U180	Nitrosopyrrolidine
U180	Pyrrole, tetrahydro-N-nitroso-
U181	5-Nitro-o-toluidine
U181	Benzenamine,2-methyl-5-nitro
U182	Paraldehyde
U182	1,3,5-Trioxane,2,4,6-trimethyl-
U183	Pentachlorobenzene
U183	Benzene, pentachloro-
U184	Pentachloroethane
U184	Ethane, pentachloro-
U185	Pentachloronitrobenzene
U185	Benzene, pentachloronitro-
U186	1,3-Pentadiene (i)
U186	1-Methylbutadiene (i)
U187	Phenacetin
U187	Acetamide, N-(4-ethoxyphenyl)
U188	Phenol
U188	Benzene,hydroxy-
U189	Phosphorus sulfide (r)
U189	Sulfur phosphide (r)
U190	Phthalic anhydride
U190	1,2-Benzenedicarboxylic acid anhydride
U191	2-Picoline
U191	Pyridine, 2-methyl-
U192	Pronamide
U192	3,5-Dichloro-N-(1,1-dimethyl-2-propynyl) benzamide
U193	1,2-Oxathiolane, 2,2-dioxide
U193	1,3-Propane sultone
U194	1-Propanamine (i,t)
U194	N-Propylamine (i,t)
U196	Pyridine
U197	p-Benzoquinone
U197	1,4-Cyclohexadienedione
U200	Reserpine
U200	Yohimban-16-carboxylic acid, 11,17-dimethoxy-18-[(3,4,5-trimethoxy-benzoyl)oxy]-, methyl ester
U201	Resorcinol
U201	1,3-Benzenediol

EPA Hazardous Waste Codes	
Code	Waste Description
U202	Saccharin and salts
U202	1,2-Benzisothiazolin-3-one,1,1-dioxide, and salts
U203	Safrole
U203	Benzene, 1,2-methylenedioxy-4-allyl-
U204	Selenious acid
U204	Selenium dioxide
U205	Selenium disulfide (r,t)
U205	Sulfur selenide (r,t)
U206	Streptozotocin
U206	D-Glucopyranose, 2-deoxy-2(3-methyl-3-nitrosoureido)-
U207	1,2,4,5-Tetrachlorobenzene
U207	Benzene, 1,2,4,5-tetrachloro-
U208	1,1,1,2-Tetrachloroethane
U208	Ethane,1,1,1,2-tetrachloro-
U209	1,1,2,2-Tetrachloroethane
U209	Ethane,1-1-2-2-tetrachloro-
U210	Tetrachloroethylene
U210	Ethene,1,1,2,2-tetrachloro
U211	Methane, tetrachloro-
U211	Carbon tetrachloride
U213	Tetrahydrofuran (i)
U213	Furan, tetrahydro- (i)
U214	Thallium(I) acetate
U214	Acetic acid, thallium(I) salt
U215	Thallium(I) carbonate
U215	Carbonic acid, dithallium(I) salt
U216	Thallium(I) chloride
U217	Thallium(I) nitrate
U218	Thioacetamide
U218	Ethanethioamide
U219	Thiourea
U219	Carbamide, thio-
U220	Toluene
U220	Benzene, methyl-
U221	Toluenediamine
U221	Diaminotoluene
U222	o-Toluidine hydrochloride

EPA Hazardous Waste Codes	
Code	Waste Description
U222	Benzenamine, 2-methyl-, hydrochloride
U223	Toluene diisocyanate (r,t)
U223	Benzene, 1,3-diisocyanatomethyl- (r,t)
U225	Methane, tribromo-
U225	Bromoform
U226	1,1,1-Trichloroethane
U226	Methylchloroform
U227	1,1,2-Trichloroethane
U227	Ethane, 1,1,2-trichloro-
U228	Trichloroethylene
U228	Trichloroethene
U234	sym-Trinitrobenzene (r,t)
U234	Benzene,1,3,5-trinitro- (r,t)
U235	1-Propanol,2,3-dibromo,phosphate (3:1)
U235	Tris(2,3-dibromopropyl) phosphate
U236	Trypan blue
U236	2,7-Naphthalenedisulfonic acid, 3,3'-[(3,3'-dimethyl-(1,1'-biphenyl)-4,4'-diyl)]-bis(azo)bis(5-amino -4-hydroxy), tetrasodium salt
U237	Uracil mustard
U237	Uracil,5-[bis(2-chloroethyl)-amino]-
U238	Ethyl carbarmate (urethan)
U238	Carbamic acid, ethyl ester
U239	Xylene (i)
U239	Benzene, dimethyl- (i,t)
U240	2,4-D, salts and esters
U240	2,4-Dichlorophenoxyacetic acid, salts, and esters
U243	1-Propene,1,1,2,3,3,3-hexachloro-
U243	Hexachloropropene
U244	Thiram
U244	Bis(dimethylthiocarbamoyl) disulfide
U246	Bromine cyanide
U246	Cyanogen bromide
U247	Ethane, 1,1,1-trichloro-2-2-bis(p-methoxyphenyl)
U247	Methoxychlor
U248	Warfarin, when present at concentrations of 0.3% or less
U248	3-(alpha-Acetonylbenzyl)-4-hydroxycoumarin and salts, when present at concentrations of 0.3% or less

EPA Hazardous Waste Codes	
Code	**Waste Description**
U249	Zinc phosphide, when present at concentrations of 10% or less
U328	2-Amino-L-methylbenzene
U328	o-Toluidine
U353	4-Amino-L-methylbenzene
U353	p-Toluidine
U359	2-Ethoxyethanol
U359	Ethylene glycol monoethyl ether

Standard Industrial Codes	
SIC Code	Industry
Agricultural Production – Crops	
0111	Wheat
0112	Rice
0115	Corn
0116	Soybeans
0119	Cash grains, nec
0131	Cotton
0132	Tobacco
0133	Sugar cane and sugar beets
0134	Irish potatoes
0139	Field crops, except cash grains, nec
0161	Vegetables and melons
0171	Berry crops
0172	Grapes
0173	Tree nuts
0174	Citrus fruits
0175	Deciduous tree fruits
0179	Fruits and tree nuts, nec
0181	Ornamental nursery products
0182	Food crops grown under cover
0191	General farms, primarily crops
Agricultural Production – Livestock	
0211	Beef cattle feedlots
0212	Beef cattle, except feedlots
0213	Hogs
0214	Sheep and goats
0219	General livestock, nec
0241	Dairy farms
0251	Broiler, fryer, and roaster chickens
0252	Chicken eggs
0253	Turkeys and turkey eggs
0254	Poultry hatcheries
0259	Poultry and eggs, nec
0271	Fur-bearing animals and rabbits
0272	Horses and other equines
0273	Animal aquaculture
0279	Animal specialties, nec

Note: nec = not elsewhere classified.

Standard Industrial Codes	
SIC Code	Industry
0291	General farms, primarily animal
Agricultural Services	
0711	Soil preparation services
0721	Crop planting and protecting
0722	Crop harvesting
0723	Crop preparation services for market
0724	Cotton ginning
0741	Veterinary services, for livestock
0742	Veterinary services, specialties
0751	Livestock services, except veterinary
0752	Animal specialty services
0761	Farm labor contractors
0762	Farm management services
0781	Landscape counseling and planning
0782	Lawn and garden services
0783	Ornamental shrub and tree services
Forestry	
0811	Timber tracts
0831	Forest products
0851	Forestry services
Fishing, Hunting, and Trapping	
0912	Finfish
0913	Shellfish
0919	Miscellaneous marine products
0921	Fish hatcheries and preserves
0971	Hunting, trapping, game propagation
Metal Mining	
1011	Iron ores
1021	Copper ores
1031	Lead and zinc ores
1041	Gold ores
1044	Silver ores
1061	Ferroalloy ores, except vanadium
1081	Metal mining services
1094	Uranium, radium, vanadium ores
1099	Metal ores, nec

Note: nec = not elsewhere classified

Standard Industrial Codes	
SIC Code	Industry
Coal Mining	
1221	Bituminous coal and lignite – surface
1222	Bituminous coal—underground
1231	Anthracite mining
1241	Coal mining services
Oil and Gas Extraction	
1311	Crude petroleum and natural gas
1321	Natural gas liquids
1381	Drilling oil and gas wells
1382	Oil and gas exploration services
1389	Oil and gas field services, nec
Nonmetallic Minerals, Except Fuels	
1411	Dimension stone
1422	Crushed and broken limestone
1423	Crushed and broken granite
1429	Crushed and broken stone, nec
1442	Construction sand and gravel
1446	Industrial sand
1455	Kaolin and ball clay
1459	Clay and related minerals, nec
1474	Potash, soda and borate minerals
1475	Phosphate rock
1479	Chemical and fertilizer mining, nec
1481	Nonmetallic minerals services
1499	Miscellaneous nonmetallic minerals, nec
General Building Contractors	
1521	Single-family housing construction
1522	Residential construction, nec
1531	Operative builders
1541	Industrial buildings and warehouses
1542	Nonresidential construction, nec
Heavy Construction, Excluding Buildings	
1611	Highway and street construction
1622	Bridge, tunnel, and elevated highway
1623	Water, sewer, and utility lines
1629	Heavy construction, nec

Note: nec = not elsewhere classified.

Standard Industrial Codes	
SIC Code	Industry
Special Trade Contractors	
1711	Plumbing, heating, air conditioning
1721	Painting and paper hanging
1731	Electrical work
1741	Masonry and other stonework
1742	Plastering, drywall, and insulation
1743	Terrazzo, tile, marble, mosaic work
1751	Carpentry work
1752	Floor laying and floor work, nec
1761	Roofing, siding, and sheet metal work
1771	Concrete work
1781	Water well drilling
1791	Structural steel erection
1793	Glass and glazing work
1794	Excavation work
1795	Wrecking and demolition work
1796	Installing building equipment, nec
1799	Special trade contractors, nec
Food and Kindred Products	
2011	Meat packing plants
2013	Sausages and other prepared meats
2015	Poultry slaughtering and processing
2021	Creamery butter
2022	Cheese, natural and processed
2023	Dry, condensed, evaporated products
2024	Ice cream and frozen desserts
2026	Fluid milk
2032	Canned specialties
2033	Canned fruits and vegetables
2034	Dehydrated fruits, vegetables, soups
2035	Pickles, sauces, and salad dressings
2037	Frozen fruits and vegetables
2038	Frozen specialties, nec
2041	Flour and other grain mill products
2043	Cereal breakfast foods
2044	Rice milling
2045	Prepared flour mixes and doughs

Note: nec = not elsewhere classified

Standard Industrial Codes	
SIC Code	Industry
2046	Wet corn milling
2047	Dog and cat food
2048	Prepared feeds, nec
2051	Bread, cake, and related products
2052	Cookies and crackers
2053	Frozen bakery products, except bread
2061	Raw cane sugar
2062	Cane sugar refining
2063	Beet sugar
2064	Candy and other confectionery products
2066	Chocolate and cocoa products
2067	Chewing gum
2068	Salted and roasted nuts and seeds
2074	Cottonseed oil mills
2075	Soybean oil mills
2076	Vegetable oil mills, nec
2077	Animal and marine fats and oils
2079	Edible fats and oils, nec
2082	Malt beverages
2083	Malt
2084	Wines, brandy, and brandy spirits
2085	Distilled and blended liquors
2086	Bottled and canned soft drinks
2087	Flavoring extracts and syrups, nec
2091	Canned and cured fish and seafood
2092	Fresh or frozen prepared fish
2095	Roasted coffee
2097	Manufactured ice
2098	Macaroni and spaghetti
2099	Food preparations, nec
Tobacco Products	
2111	Cigarettes
2121	Cigars
2131	Chewing and smoking tobacco
2141	Tobacco stemming and redrying

Note: nec = not elsewhere classified.

Standard Industrial Codes	
SIC Code	Industry
Textile Mill Products	
2211	Broadwoven fabric mills, cotton
2221	Broadwoven fabric mills, man-made
2231	Broadwoven fabric mills, wool
2241	Narrow fabric mills
2251	Women's hosiery, except socks
2252	Hosiery, nec
2253	Knit outerwear mills
2254	Knit underwear mills
2257	Weft knit fabric mills
2258	Lace and warp knit fabric mills
2259	Knitting mills, nec
2261	Finishing plants, cotton
2262	Finishing plants, man-made
2269	Finishing plants, nec
2273	Carpets and rugs
2281	Yarn spinning mills
2282	Throwing and winding mills
2284	Thread mills
2295	Coated fabrics, not rubberized
2296	Tire cord and fabrics
2297	Nonwoven fabrics
2298	Cordage and twine
2299	Textile goods, nec
Apparel and Other Textile Products	
2311	Men's and boys' suits and coats
2321	Men's and boys' shirts
2322	Men's and boys' underwear and nightwear
2323	Men's and boys' neckwear
2325	Men's and boys' trousers and slacks
2326	Men's and boys' work clothing
2329	Men's and boys' clothing, nec
2331	Women's and misses' blouses and shirts
2335	Women's, juniors' and misses' dresses
2337	Women's and misses' suits and coats
2339	Women's and misses' outerwear, nec
2341	Women's and children's underwear

Note: nec = not elsewhere classified

SIC Code	Industry
2342	Bras, girdles, and allied garments
2353	Hats, caps, and millinery
2361	Girls' and children's dresses, blouses
2369	Girls' and children's outerwear, nec
2371	Fur goods
2381	Fabric dress and work gloves
2384	Robes and dressing gowns
2385	Waterproof outerwear
2386	Leather and sheep lined clothing
2387	Apparel belts
2389	Apparel and accessories, nec
2391	Curtains and draperies
2392	House furnishings, nec
2393	Textile bags
2394	Canvas and related products
2395	Pleating and stitching
2396	Automotive and apparel trimmings
2397	Schiffli machine embroideries
2399	Fabricated textile products, nec
Lumber and Wood Products	
2411	Logging
2421	Sawmills and planing mills, general
2426	Hardwood dimension and flooring mills
2429	Special product sawmills, nec
2431	Millwork
2434	Wood kitchen cabinets
2435	Hardwood veneer and plywood
2436	Softwood veneer and plywood
2439	Structural wood members, nec
2441	Nailed wood boxes and shook
2448	Wood pallets and skids
2449	Wood containers, nec
2451	Mobile homes
2452	Prefabricated wood buildings
2491	Wood preserving
2493	Reconstituted wood products
2499	Wood products, nec

Standard Industrial Codes

Note: nec = not elsewhere classified.

Standard Industrial Codes	
SIC Code	Industry
Furniture and Fixtures	
2511	Wood household furniture
2512	Upholstered household furniture
2514	Metal household furniture
2515	Mattresses and bedsprings
2517	Wood TV and radio cabinets
2519	Household furniture, nec
2521	Wood office furniture
2522	Office furniture, except wood
2531	Public building and related furniture
2541	Wood partitions and fixtures
2542	Partitions and fixtures, except wood
2591	Drapery hardware and blinds and shades
2599	Furniture and fixtures, nec
Paper and Allied Products	
2611	Pulp mills
2621	Paper mills
2631	Paperboard mills
2652	Set-up paperboard boxes
2653	Corrugated and solid fiber boxes
2655	Fiber cans, drums, and similar products
2656	Sanitary food containers
2657	Folding paperboard boxes
2671	Paper coated and laminated, packaging
2672	Paper coated and laminated, nec
2673	Bags – plastics, laminated and coated
2674	Bags – uncoated paper and multiwall
2675	Die-cut paper and board
2676	Sanitary paper products
2677	Envelopes
2678	Stationery products
2679	Converted paper products, nec
Printing and Publishing	
2711	Newspapers
2721	Periodicals
2731	Book publishing

Note: nec = not elsewhere classified

Standard Industrial Codes	
SIC Code	Industry
2732	Book printing
2741	Miscellaneous publishing
2752	Commercial printing, lithographic
2754	Commercial printing, gravure
2759	Commercial printing, nec
2761	Manifold business forms
2771	Greeting cards
2782	Blankbooks and looseleaf binders
2789	Bookbinding and related work
2791	Typesetting
2796	Plate making services
Chemicals and Allied Products	
2812	Alkalies and chlorine
2813	Industrial gases
2816	Inorganic pigments
2819	Industrial inorganic chemicals, nec
2821	Plastics materials and resins
2822	Synthetic rubber
2823	Cellulosic man-made fibers
2824	Organic fibers, noncellulosic
2833	Medicinals and botanicals
2834	Pharmaceutical preparations
2835	Diagnostic substances
2836	Biological products, except diagnostic
2841	Soap and other detergents
2842	Polishes and sanitation goods
2843	Surface active agents
2844	Toilet preparations
2851	Paints and allied products
2861	Gum and wood chemicals
2865	Cyclic crudes and intermediates
2869	Industrial organic chemicals, nec
2873	Nitrogenous fertilizers
2874	Phosphatic fertilizers
2875	Fertilizers, mixing only
2879	Agricultural chemicals, nec
2891	Adhesives and sealants

Note: nec = not elsewhere classified.

Standard Industrial Codes	
SIC Code	Industry
2892	Explosives
2893	Printing ink
2895	Carbon black
2899	Chemical preparations, nec
Petroleum and Coal Products	
2911	Petroleum refining
2951	Asphalt paving mixtures and blocks
2952	Asphalt felts and coatings
2992	Lubricating oils and greases
2999	Petroleum and coal products, nec
Rubber and Miscellaneous Plastic Products	
3011	Tires and inner tubes
3021	Rubber and plastics footwear
3052	Rubber and plastics hose and belting
3053	Gaskets, packing and sealing devices
3061	Mechanical rubber goods
3069	Fabricated rubber products, nec
3081	Unsupported plastics, film and sheet
3082	Unsupported plastics, profile shapes
3083	Laminated plastics, plate and sheet
3084	Plastics, pipe
3085	Plastics, bottles
3086	Plastics, foam products
3087	Custom compound purchased resins
3088	Plastics, plumbing futures
3089	Plastics products, nec
Leather and Leather Products	
3111	Leather tanning and finishing
3131	Footwear, cut stock
3142	House slippers
3143	Men's footwear, except athletic
3144	Women's footwear, except athletic
3149	Footwear, except rubber, nec
3151	Leather gloves and mittens
3161	Luggage
3171	Women's handbags and purses

Note: nec = not elsewhere classified

Standard Industrial Codes	
SIC Code	Industry
3172	Personal leather goods, nec
3199	Leather goods, nec
Stone, Clay, and Glass Products	
3211	Flat glass
3221	Glass containers
3229	Pressed and blown glass, nec
3231	Products of purchased glass
3241	Cement, hydraulic
3251	Brick and structural clay tile
3253	Ceramic wall and floor tile
3255	Clay refractories
3259	Structural clay products, nec
3261	Vitreous plumbing fixtures
3262	Vitreous china table and kitchenware
3263	Semivitreous table and kitchenware
3264	Porcelain electrical supplies
3269	Pottery products, nec
3271	Concrete block and brick
3272	Concrete products, nec
3273	Ready-mixed concrete
3274	Lime
3275	Gypsum products
3281	Cut stone and stone products
3291	Abrasive products
3292	Asbestos products
3295	Minerals, ground or treated
3296	Mineral wool
3297	Nonclay refractories
3299	Nonmetallic mineral products, nec
Primary Metal Industries	
3312	Blast furnaces and steel mills
3313	Electrometallurgical products
3315	Steel wire and related products
3316	Cold finishing of steel shapes
3317	Steel pipe and tubes
3321	Gray and ductile iron foundries

Note: nec = not elsewhere classified.

Standard Industrial Codes	
SIC Code	Industry
3322	Malleable iron foundries
3324	Steel investment foundries
3325	Steel foundries, nec
3331	Primary copper
3334	Primary aluminum
3339	Primary nonferrous metals, nec
3341	Secondary nonferrous metals
3351	Copper rolling and drawing
3353	Aluminum sheet, plate, and foil
3354	Aluminum extruded products
3355	Aluminum rolling and drawing, nec
3356	Nonferrous rolling and drawing, nec
3357	Nonferrous wire drawing and insulating
3363	Aluminum die-castings
3364	Nonferrous die-castings, except aluminum
3365	Aluminum foundries
3366	Copper foundries
3369	Nonferrous foundries, nec
3398	Metal heat treating
3399	Primary metal products, nec
Fabricated Metal Products	
3411	Metal cans
3412	Metal barrels, drums, and pails
3421	Cutlery
3423	Hand and edge tools, nec
3425	Saw blades and handsaws
3429	Hardware, nec
3431	Metal sanitary ware
3432	Plumbing fixture fittings and trim
3433	Heating equipment, except electric
3441	Fabricated structural metal
3442	Metal doors, sash, and trim
3443	Fabricated plate work (boiler shops)
3444	Sheet metal work
3446	Architectural metal work
3448	Prefabricated metal buildings
3449	Miscellaneous metal work

Note: nec = not elsewhere classified

Standard Industrial Codes	
SIC Code	Industry
3451	Screw machine products
3452	Bolts, nuts, rivets, and washers
3462	Iron and steel forgings
3463	Nonferrous forgings
3465	Automotive stampings
3466	Crowns and closures
3469	Metal stampings, nec
3471	Plating and polishing
3479	Metal coating and allied services
3482	Small arms ammunition
3483	Ammunition, except for small arms, nec
3484	Small arms
3489	Ordnance and accessories, nec
3491	Industrial valves
3492	Fluid power valves and hose fittings
3493	Steel springs, except wire
3494	Valves and pipe fittings, nec
3495	Wire springs
3496	Miscellaneous fabricated wire products
3497	Metal foil and leaf
3498	Fabricated pipe and fittings
3499	Fabricated metal products, nec
Industrial Machinery and Equipment	
3511	Turbines and turbine generator sets
3519	Internal combustion engines, nec
3523	Farm machinery and equipment
3524	Lawn and garden equipment
3531	Construction machinery
3532	Mining machinery
3533	Oil and gas field machinery
3534	Elevators and moving stairways
3535	Conveyors and conveying equipment
3536	Hoists, cranes, and monorails
3537	Industrial trucks and tractors
3541	Machine tools, metal cutting types
3542	Machine tools, metal forming types
3543	Industrial patterns

Note: nec = not elsewhere classified.

Standard Industrial Codes	
SIC Code	Industry
3544	Special dies, tools, jigs, and fixture
3545	Machine tool accessories
3546	Power driven hand tools
3547	Rolling mill machinery
3548	Welding apparatus
3549	Metalworking machinery, nec
3552	Textile machinery
3553	Woodworking machinery
3554	Paper industries machinery
3555	Printing trades machinery
3556	Food products machinery
3559	Special industry machinery, nec
3561	Pumps and pumping equipment
3562	Ball and roller bearings
3563	Air and gas compressors
3564	Blowers and fans
3565	Packaging machinery
3566	Speed changers, drives, and gears
3567	Industrial furnaces and ovens
3568	Power transmission equipment, nec
3569	General industrial machinery, nec
3571	Electronic computers
3572	Computer storage devices
3575	Computer terminals
3577	Computer peripheral equipment, nec
3578	Calculating and accounting equipment
3579	Office machines, nec
3581	Automatic vending machines
3582	Commercial laundry equipment
3585	Refrigeration and heating equipment
3586	Measuring and dispensing pumps
3589	Service industry machinery, nec
3592	Carburetors, pistons, rings, valves
3593	Fluid power cylinders and actuators
3594	Fluid power pumps and motors
3596	Scales and balances, except laboratory
3599	Industrial machinery, nec

Note: nec = not elsewhere classified

Standard Industrial Codes	
SIC Code	Industry
Electronic and Other Electric Equipment	
3612	Transformers, except electronic
3613	Switchgear and switchboard apparatus
3621	Motors and generators
3624	Carbon and graphite products
3625	Relays and industrial controls
3629	Electrical industrial apparatus, nec
3631	Household cooking equipment
3632	Household refrigerators and freezers
3633	Household laundry equipment
3634	Electric housewares and fans
3635	Household vacuum cleaners
3639	Household appliances, nec
3641	Electric lamps
3643	Current-carrying wiring devices
3644	Noncurrent-carrying wiring devices
3645	Residential lighting fixtures
3646	Commercial lighting fixtures
3647	Vehicular lighting equipment
3648	Lighting equipment, nec
3651	Household audio and video equipment
3652	Prerecorded records and tapes
3661	Telephone and telegraph apparatus
3663	Radio and TV communication equipment
3669	Communications equipment, nec
3671	Electron tubes
3672	Printed circuit boards
3674	Semiconductors and related devices
3675	Electronic capacitors
3676	Electronic resistors
3677	Electronic coils and transformers
3678	Electronic connectors
3679	Electronic components, nec
3691	Storage batteries
3692	Primary batteries, dry and wet
3694	Engine electrical equipment
3695	Magnetic and optical recording media

Note: nec = not elsewhere classified.

Standard Industrial Codes	
SIC Code	**Industry**
3699	Electrical equipment and supplies, nec
Transportation Equipment	
3711	Motor vehicles and car bodies
3713	Truck and bus bodies
3714	Motor vehicle parts and accessories
3715	Truck trailers
3716	Motor homes
3721	Aircraft
3724	Aircraft engines and engine parts
3728	Aircraft parts and equipment, nec
3731	Ship building and repairing
3732	Boat building and repairing
3743	Railroad equipment
3751	Motorcycles, bicycles, and parts
3761	Guided missiles and space vehicles
3764	Space propulsion units and parts
3769	Space vehicle equipment, nec
3792	Travel trailers and campers
3795	Tanks and tank components
3799	Transportation equipment, nec
Instruments and Related Products	
3812	Search and navigation equipment
3821	Laboratory apparatus and furniture
3822	Environmental controls
3823	Process control instruments
3824	Fluid meters and counting devices
3825	Instruments to measure electricity
3826	Analytical instruments
3827	Optical instruments and lenses
3829	Measuring and controlling devices, nec
3841	Surgical and medical instruments
3842	Surgical appliances and supplies
3843	Dental equipment and supplies
3844	X-ray apparatus and tubes
3845	Electromedical equipment
3851	Ophthalmic goods

Note: nec = not elsewhere classified

Standard Industrial Codes	
SIC Code	Industry
3861	Photographic equipment and supplies
3873	Watches, clocks, watch cases, and parts
Miscellaneous Manufacturing Industries	
3911	Jewelry, precious metal
3914	Silverware and plated ware
3915	Jewelers' materials and lapidary work
3931	Musical instruments
3942	Dolls and stuffed toys
3944	Games, toys, and children's vehicles
3949	Sporting and athletic goods, nec
3951	Pens and mechanical pencils
3952	Lead pencils and art goods
3953	Marking devices
3955	Carbon paper and inked ribbons
3961	Costume jewelry
3965	Fasteners, buttons, needles, and pins
3991	Brooms and brushes
3993	Signs and advertising specialties
3995	Burial caskets
3996	Hard surface floor coverings, nec
3999	Manufacturing industries, nec
Railroad Transportation	
4011	Railroads, line-haul operating
4013	Switching and terminal devices
Local and Interurban Passenger Transit	
4111	Local and suburban transit
4119	Local passenger transportation, nec
4121	Taxicabs
4131	Intercity and rural bus transportation
4141	Local bus charter service
4142	Bus charter service, except local
4151	School buses
4173	Bus terminal and service facilities
Trucking and Warehousing	
4212	Local trucking, without storage
4213	Trucking, except local

Note: nec = not elsewhere classified.

Standard Industrial Codes	
SIC Code	Industry
4214	Local trucking with storage
4215	Courier services, except by air
4221	Farm product warehousing and storage
4222	Refrigerated warehousing and storage
4225	General warehousing and storage
4226	Special warehousing and storage, nec
4231	Trucking terminal facilities
U.S. Postal Service	
4311	U.S. Postal Service
Water Transportation	
4412	Deep sea foreign transportation of freight
4424	Deep sea domestic transportation of freight
4432	Freight transportation, on the Great Lakes
4449	Water transportation of freight, nec
4481	Deep sea passenger transportation, except ferry
4482	Ferries
4489	Water passenger transportation, nec
4491	Marine cargo handling
4492	Towing and tugboat service
4493	Marinas
4499	Water transportation services, nec
Transportation by Air	
4512	Air transportation, scheduled
4513	Air courier services
4522	Air transportation, nonscheduled
4581	Airports, flying fields, and services
Pipelines, Except Natural Gas	
4612	Crude petroleum pipelines
4613	Refined petroleum pipelines
4619	Pipelines, nec
Transportation Services	
4724	Travel agencies
4725	Tour operators
4729	Passenger transportation arrangement, nec
4731	Freight transportation arrangement
4741	Rental of railroad cars

Note: nec = not elsewhere classified

Standard Industrial Codes	
SIC Code	Industry
4783	Packing and crating
4785	Inspection and fixed facilities
4789	Transportation services, nec
Communications	
4812	Radio-telephone communications
4813	Telephone communications, except radio
4822	Telegraph and other communications
4832	Radio broadcasting stations
4833	Television broadcasting stations
4841	Cable and other pay TV services
4899	Communication services, nec
Electric, Gas, and Sanitary Services	
4911	Electric services
4922	Natural gas transmission
4923	Gas transmission and distribution
4924	Natural gas distribution
4925	Gas production and/or distribution
4931	Electric and other services combined
4932	Gas and other services combined
4939	Combination utilities, nec
4941	Water supply
4952	Sewerage systems
4953	Refuse systems
4959	Sanitary services, nec
4961	Steam and air conditioning supply
4971	Irrigation systems
Wholesale Trade, Durable Goods	
5012	Automobiles and other motor vehicles
5013	Motor vehicle supplies and new parts
5014	Tires and tubes
5015	Motor vehicle parts, used
5021	Furniture
5023	Home furnishings
5031	Lumber, plywood, and millwork
5032	Brick, stone, and related materials
5033	Roofing, siding, and insulation

Note: nec = not elsewhere classified.

Standard Industrial Codes	
SIC Code	Industry
5039	Construction materials, nec
5043	Photographic equipment and supplies
5044	Office equipment
5045	Computers, peripherals, and software
5046	Commercial equipment, nec
5047	Medicinal and hospital equipment
5048	Ophthalmic goods
5049	Professional equipment, nec
5051	Metals service centers and offices
5052	Coal and other minerals and ores
5063	Electrical apparatus and equipment
5064	Electrical appliances, TV and radios
5065	Electronic parts and equipment
5072	Hardware
5074	Plumbing and hydronic heating supplies
5075	Warm air heating and air conditioning
5078	Refrigeration equipment and supplies
5082	Construction and mining machinery
5083	Farm and garden machinery
5084	Industrial machinery and equipment
5085	Industrial supplies
5087	Service establishment equipment
5088	Transportation equipment and supplies
5091	Sporting and recreational goods
5092	Toys and hobby goods and supplies
5093	Scrap and waste materials
5094	Jewelry and precious stones
5099	Durable goods, nec
Wholesale Trade, Nondurable Goods	
5111	Printing and writing paper
5112	Stationery and office supplies
5113	Industrial and personal service paper
5122	Drugs, proprietaries, and sundries
5131	Piece goods and notions
5136	Men's and boys' clothing
5137	Women's and children's clothing
5139	Footwear

Note: nec = not elsewhere classified

Standard Industrial Codes

SIC Code	Industry
5141	Groceries, general line
5142	Packaged frozen foods
5143	Dairy products, except dried or canned
5144	Poultry and poultry products
5145	Confectionery
5146	Fish and seafoods
5147	Meats and meat products
5148	Fresh fruits and vegetables
5149	Groceries and related products, nec
5153	Grain and field beans
5154	Livestock
5159	Farm-product raw materials, nec
5162	Plastics materials and basic shapes
5169	Chemicals and allied products, nec
5171	Petroleum bulk stations and terminals
5172	Petroleum products, nec
5181	Beer and ale
5182	Wines and distilled beverages
5191	Farm supplies
5192	Books, periodicals, and newspapers
5193	Flowers and florists' supplies
5194	Tobacco and tobacco products
5198	Paints, varnishes, and supplies
5199	Nondurable goods, nec
Building Materials and Garden Supplies	
5211	Lumber and other building materials
5231	Paint, glass, and wallpaper stores
5251	Hardware stores
5261	Retail nurseries and gardens
5271	Mobile home dealers
General Merchandise Stores	
5311	Department stores
5331	Variety stores
5399	Miscellaneous general merchandise stores
Food Stores	
5411	Grocery stores

Note: nec = not elsewhere classified.

Standard Industrial Codes	
SIC Code	Industry
5421	Meat and fish markets
5431	Fruit and vegetable markets
5441	Candy, nut, and confectionery stores
5451	Dairy products stores
5461	Retail bakers
5499	Miscellaneous food stores
Automotive Dealers and Service Stations	
5511	New and used car dealers
5521	Used car dealers
5531	Auto and home supply stores
5541	Gasoline service stations
5551	Boat dealers
5561	Recreational vehicle dealers
5571	Motorcycle dealers
5599	Automotive dealers, nec
Apparel and Accessory Stores	
5611	Men's and boys' clothing stores
5621	Women's clothing stores
5632	Women's accessory and specialty stores
5641	Children's and infants' wear stores
5651	Family clothing stores
5661	Shoe stores
5699	Miscellaneous apparel and accessory stores
Furniture and Home Furnishings Stores	
5712	Furniture stores
5713	Floor covering stores
5714	Drapery and upholstery stores
5719	Miscellaneous home furnishings stores
5722	Household appliance stores
5731	Radio, TV, and electronic stores
5734	Computer and software stores
5735	Record and prerecorded tape stores
5736	Musical instruments stores
Eating and Drinking Places	
5812	Eating places
5813	Drinking places

Note: nec = not elsewhere classified

Standard Industrial Codes	
SIC Code	**Industry**
Miscellaneous Retail	
5912	Drugstores and proprietary stores
5921	Liquor stores
5932	Used merchandise stores
5941	Sporting goods and bicycle shops
5942	Bookstores
5943	Stationery stores
5944	Jewelry stores
5945	Hobby, toy, and game shops
5946	Camera and photographic supply stores
5947	Gift, novelty, and souvenir shops
5948	Luggage and leather goods stores
5949	Sewing, needlework, and piece goods
5961	Catalog and mail order houses
5962	Merchandising machine operators
5963	Direct selling organizations
5983	Fuel oil dealers
5984	Liquefied petroleum gas dealers
5989	Fuel dealers, nec
5992	Florists
5993	Cigar stores and stands
5994	News dealers and newsstands
5995	Optical goods stores
5999	Miscellaneous retail stores, nec
Depository Institutions	
6011	Federal Reserve banks
6019	Central reserve depository, nec
6021	National commercial banks
6022	State commercial banks
6029	Commercial banks, nec
6035	Federal savings institutions
6036	Savings institutions, except federal
6061	Federal credit unions
6062	State credit unions
6081	Foreign banks and branches and agencies
6082	Foreign trade and international banks
6091	Nondeposit trust facilities

Note: nec = not elsewhere classified.

Standard Industrial Codes	
SIC Code	**Industry**
6099	Functions related to deposit banking
Nondepository Institutions	
6111	Federal and federally-sponsored credit
6141	Personal credit institutions
6153	Short-term business credit
6159	Miscellaneous business credit institutions
6162	Mortgage bankers and correspondents
6163	Loan brokers
Security and Commodity Brokers	
6211	Security brokers and dealers
6221	Commodity contracts brokers, dealers
6231	Security and commodity exchanges
6282	Investment advice
6289	Security and commodity services, nec
Insurance Carriers	
6311	Life insurance
6321	Accident and health insurance
6324	Hospital and medical service plans
6331	Fire, marine, and casualty insurance
6351	Surety insurance
6361	Title insurance
6371	Pension, health, and welfare funds
6399	Insurance carriers, nec
Insurance Agents, Brokers, and Service	
6411	Insurance agents, brokers, and service
Real Estate	
6512	Nonresidential building operators
6513	Apartment building operators
6514	Dwelling operators, except apartments
6515	Mobile home site operators
6517	Railroad property lessors
6519	Real property lessors, nec
6531	Real estate agents and managers
6541	Title abstract offices
6552	Subdividers and developers, nec

Note: nec = not elsewhere classified

Standard Industrial Codes	
SIC Code	Industry
6553	Cemetery subdividers and developers
Holding and Other Investment Offices	
6712	Bank holding companies
6719	Holding companies, nec
6722	Management investment, open-end
6726	Investment offices, nec
6732	Educational, religious, etc. trusts
6733	Trusts, nec
6792	Oil royalty traders
6794	Patent owners and lessors
6798	Real estate investment trusts
6799	Investors, nec
Hotels and Other Lodging Places	
7011	Hotels and motels
7021	Rooming and boarding houses
7032	Sporting and recreational camps
7033	Trailer parks and campsites
7041	Membership-basis organization hotels
Personal Services	
7211	Power laundries, family and commercial
7212	Garment pressing and cleaners' agents
7213	Linen supply
7215	Coin-operated laundries and cleaning
7216	Dry cleaning plants, except rug
7217	Carpet and upholstery cleaning
7218	Industrial launderers
7219	Laundry and garment services, nec
7221	Photographic studios, portrait
7231	Beauty shops
7241	Barber shops
7251	Shoe repair and shoeshine shops
7261	Funeral service and crematories
7291	Tax return preparation services
7299	Miscellaneous personal services, nec
Business Services	
7311	Advertising agencies

Note: nec = not elsewhere classified.

Standard Industrial Codes	
SIC Code	Industry
7312	Outdoor advertising services
7313	Radio, TV, publisher representatives
7319	Advertising, nec
7322	Adjustment and collection services
7323	Credit reporting services
7331	Direct mail advertising services
7334	Photocopying and duplicating services
7335	Commercial photography
7336	Commercial art and graphic design
7338	Secretarial and court reporting
7342	Disinfecting and pest control services
7349	Building maintenance services, nec
7352	Medical equipment rental
7353	Heavy construction equipment rental
7359	Equipment rental and leasing, nec
7361	Employment agencies
7363	Help supply services
7371	Computer programming services
7372	Prepackaged software
7373	Computer integrated systems design
7374	Data processing services
7375	Information retrieval services
7376	Computer facilities management
7377	Computer rental and leasing
7378	Computer maintenance and repair
7379	Computer related services, nec
7381	Detective and armored car services
7382	Security systems services
7383	News syndicates
7384	Photofinishing laboratories
7389	Business services, nec
Automotive Repair, Services, and Parking	
7513	Truck rental and leasing, no drivers
7514	Passenger car rental
7515	Passenger car leasing
7519	Utility trailer rental
7521	Automobile parking

Note: nec = not elsewhere classified

Standard Industrial Codes	
SIC Code	Industry
7532	Top and body repair and paint shops
7533	Auto exhaust system repair shops
7534	Tire retreading and repair shops
7536	Automotive glass replacement shops
7537	Automotive transmission repair shops
7538	General automotive repair shops
7539	Automotive repair shops, nec
7542	Car washes
7549	Automotive services, nec
Miscellaneous Repair Services	
7622	Radio and television repair
7623	Refrigeration service and repair
7629	Electrical repair shops, nec
7631	Watch, clock, and jewelry repair
7641	Reupholstery and furniture repair
7692	Welding repair
7694	Armature rewinding shops
7699	Repair services, nec
Motion Pictures	
7812	Motion picture and video production
7819	Services allied to motion pictures
7822	Motion picture and tape distribution
7829	Motion picture distribution services
7832	Motion picture theaters except drive-in
7833	Drive-in motion picture theaters
7841	Video tape rental
Amusement and Recreation Services	
7911	Dance studios, schools, and halls
7922	Theatrical producers and services
7929	Entertainers and entertainment groups
7933	Bowling centers
7941	Sports clubs, managers, and promoters
7948	Racing, including track operation
7991	Physical fitness facilities
7992	Public golf courses
7993	Coin-operated amusement devices

Note: nec = not elsewhere classified.

Standard Industrial Codes	
SIC Code	Industry
7996	Amusement parks
7997	Membership sports and recreation clubs
7999	Amusement and recreation, nec
Health Services	
8011	Offices and clinics of medical doctors
8021	Offices and clinics of dentists
8031	Offices of osteopathic physicians
8041	Offices and clinics of chiropractors
8042	Offices and clinics of optometrists
8043	Offices and clinics of podiatrists
8049	Offices of health practitioners, nec
8051	Skilled nurse care facilities
8052	Intermediate care facilities
8059	Nursing and personal care, nec
8062	General medical and surgical hospitals
8063	Psychiatric hospitals
8069	Specialty hospitals, except psychiatric
8071	Medical laboratories
8072	Dental laboratories
8082	Home health care services
8092	Kidney dialysis centers
8093	Specialty outpatient clinics, nec
8099	Health and allied services, nec
Legal Services	
8111	Legal services
Educational Services	
8211	Elementary and secondary schools
8221	Colleges and universities
8222	Junior colleges
8231	Libraries
8243	Data processing schools
8244	Business and secretarial schools
8249	Vocational schools, nec
8299	Schools and educational services, nec
Social Services	
8322	Individual and family services

Note: nec = not elsewhere classified

SIC Code	Industry
8331	Job training and related services
8351	Child day care services
8361	Residential care
8399	Social services, nec
Museums, Botanical, Zoological Gardens	
8412	Museums and art galleries
8422	Botanical and zoological gardens
Membership Organizations	
8611	Business associations
8621	Professional organizations
8631	Labor organizations
8641	Civic and social associations
8651	Political organizations
8661	Religious organizations
8699	Membership organizations, nec
Engineering and Management Services	
8711	Engineering services
8712	Architectural services
8713	Surveying services
8721	Accounting, auditing, and bookkeeping
8731	Commercial physical research
8732	Commercial nonphysical research
8733	Noncommercial research organizations
8734	Testing laboratories
8741	Management services
8742	Management consulting services
8743	Public relations services
8744	Facilities support services
8748	Business consulting, nec
Private Households	
8811	Private households
Services, nec	
8999	Services, nec
Executive, Legislative, and General	
9111	Executive offices

Standard Industrial Codes

Note: nec = not elsewhere classified.

SIC Code	Industry
9121	Legislative bodies
9131	Executive and legislative combined
9199	General government, nec
Justice, Public Order, and Safety	
9211	Courts
9221	Police protection
9222	Legal counsel and prosecution
9223	Correctional institutions
9224	Fire protection
9229	Public order and safety, nec
Finance, Taxation, and Monetary Policy	
9311	Finance, taxation, and monetary policy
Administration of Human Resources	
9411	Administration of educational programs
9431	Administration of public health programs
9441	Administration of social and manpower programs
9451	Administration of veterans' affairs
Environmental Quality, and Housing	
9511	Air, water, and solid waste management
9512	Land, mineral, wildlife conservation
9531	Housing programs
9532	Urban and community development
Administration of Economic Programs	
9611	Administration of general economic programs
9621	Regulation, administration of transportation
9631	Regulation, administration of utilities
9641	Regulation of agricultural marketing
9651	Regulation of miscellaneous commercial sectors
9661	Space research and technology
National Security and International Affairs	
9711	National security
9721	International affairs
Nonclassifiable Establishments	
9999	Nonclassifiable establishments

Standard Industrial Codes

Note: nec = not elsewhere classified

State Hazardous Waste Management Agencies

Alabama
Alabama Department of Environmental Management
Land Division
1751 Federal Drive
Montgomery, Alabama 36130
(205) 271-7730

Alaska
Department of Environmental Conservation
P.O. Box 0
Juneau, Alaska 99811
Program Manager:
(907) 465-2666
Northern Regional Office (Fairbanks):
(907) 452-1714
South-Central Regional Office (Anchorage):
(907) 274-2533
Southeast Regional Office (Juneau):
(907) 789-3151

American Samoa
Environmental Quality Commission
Government of American Samoa
Pago Pago, American Samoa 96799
Overseas Operator
(Commercial Call (684) 663-4116)

Arizona
Arizona Department of Health Services
Office of Waste and Water Quality
2005 North Central Avenue
Room 304
Phoenix, Arizona 85004
Hazardous Waste Management: (602) 257-6801

Arkansas
Department of Pollution Control Ecology
Hazardous Waste Division
P.O. Box 9583
8001 National Drive
Little Rock, Arkansas 72219
(501) 562-7444

California
Department of Health Services
Toxic Substance Control Division
714 P Street
Room 1253
Sacramento, California 95814
(916) 324-1826

State Waste Resources Control Board
Division of Water Quality
P.O. Box 100
Sacramento, California 95801
(916) 322-2867

Colorado
Colorado Department of Health
Waste Management Division
4210 E. 11th Avenue
Denver, Colorado 80220
(303) 320-8333 Ext. 4364

Connecticut
Department of Environmental Protection
Hazardous Waste Management Section
State Office Building
165 Capitol Avenue
Hartford, Connecticut 06106
(203) 566-8843, 8844

Connecticut Resource Recovery Authority
179 Allyn Street, Suite 603
Professional Building
Hartford, Connecticut 06103
(203) 549-6390

Delaware
Department of Natural Resources and Environmental Control
Waste Management Section
P.O. Box 1401
Dover, Delaware 19903
(302) 736-4781

District of Columbia
Department of Consumer and Regulatory Affairs
Pesticides and Hazardous Waste Materials Division
Room 114
5010 Overlook Avenue, S.W.
Washington, D.C. 20032
(202) 767-8414

Florida
Department of Environmental Regulation
Solid and Hazardous Waste Section
Twin Towers Office Building
2600 Blair Stone Road
Tallahassee, Florida 32301
RE: SQG's
(904) 488-0300

Georgia
Georgia Environmental Protection Division
Hazardous Waste Management Program
Land Protection Branch
Floyd Towers East, Suite 1154
205 Butler Street, S.E.
Atlanta, Georgia 30334
(404) 669-3927
Toll Free: (800) 334-2373

Guam
Guam Environmental Protection Agency
P.O. Box 2999
Agana, Guam 96910
Overseas Operator
(Commercial Call (671) 646-7579)

State Hazardous Waste Management Agencies

Hawaii
Department of Health
Environmental Health Division
P.O. Box 3378
Honolulu, Hawaii 96801
(808) 548-4383

Idaho
Department of Health and Welfare
Bureau of Hazardous Materials
450 West State Street
Boise, Idaho 83720
(208) 334-5879

Illinois
Environmental Protection Agency
Division of Land Pollution Control
2200 Chruchill Road, #24
Springfield, Illinois 62706
(217) 782-6761

Indiana
Department of Environmental Management
Office of Solid and Hazardous Waste
105 South Meridian
Indianapolis, Indiana 46225
(317) 232-4535

Iowa
U.S. EPA Region VII
Hazardous Materials Branch
726 Minnesota Avenue
Kansas City, Kansas 66101
(913) 236-2888
Iowa RCRA Toll Free: (800) 223-0425

Kansas
Department of Health and Environment
Bureau of Waste Management
Forbes Field, Building 321
Topeka, Kansas 66620
(913) 862-9360 Ext. 292

Kentucky
Natural Resources and Environmental Protection Cabinet
Division of Waste Management
18 Reilly Road
Frankfort, Kentucky 40601
(502) 564-6716

Louisiana
Department of Environmental Quality
Hazardous Waste Division
P.O. Box 44307
Baton Rouge, Louisiana 70804
(504) 342-1227

Maine
Department of Environmental Protection
Bureau of Oil and Hazardous Materials Control
State House Station #17
Augusta, Maine 04333
(207) 289-2651

Maryland
Department of Health and Mental Hygiene
Maryland Waste Management Administration
Office of Environmental Programs
201 West Preston Street, Room A3
Baltimore, Maryland 21201
(301) 631-3343

Massachusetts
Department of Environmental Quality Engineering
Division of Solid and Hazardous Waste
One Winter Street, 5th Floor
Boston, Massachusetts 02108
(617) 292-5851

Michigan
Michigan Department of Natural Resources
Hazardous Waste Division
Waste Evaluation Unit
Box 30028
Lansing, Michigan 48909
(517) 373-2730

Minnesota
Pollution Control Agency
Solid and Hazardous Waste Division
520 Lafayette Road
St. Paul, Minnesota 55155
(612) 296-6300

Mississippi
Department of Natural Resources
Division of Solid and Hazardous Waste Management
P.O. Box 10385
Jackson, Mississippi 39209
(601) 961-5062

Missouri
Department of Natural Resources
Waste Management Program
P.O. Box 176
Jefferson City, Missouri 65102
(314) 751-3176
Missouri Hotline: (800) 334-6946

Montana
Department of Health and Environmental Sciences
Solid and Hazardous Waste Bureau
Cogswell Building, B-201
Helena, Montana 59620
(406) 444-2821

Nebraska
Department of Environmental Control
Hazardous Waste Management Section
P.O. Box 94877
State House Station
Lincoln, Nebraska 68509
(402) 471-2186

Nevada
Division of Environmental Protection
Waste Management Program
Capitol Complex
Carson City, Nevada 89710
(702) 885-5872

State Hazardous Waste Management Agencies

New Hampshire
Department of Health and Human Services
Division of Public Health Services
Office of Waste Management
Health and Welfare Building
Hazen Drive
Concord, New Hampshire 03301-6527
(603) 271-2944

New Jersey
Department of Environmental Protection
Division of Waste Management
32 East Hanover Street, CN-028
Trenton, New Jersey 08625
Hazardous Waste Advisement Program: (609) 292-8341

New Mexico
Environmental Improvement Division
Ground Water And Hazardous Waste Bureau
Hazardous Waste Section
P.O. Box 968
Santa Fe, New Mexico 87504-0968
(505) 827-2922

New York
Department of Environmental Conservation
Bureau of Hazardous Waste Operations
50 Wolf Road, Room 209
Albany, New York 12233
(518) 457-0530
SQG Hotline: (800) 631-0666

North Carolina
Department of Human Resources
Solid and Hazardous Waste Management Branch
P.O. Box 2091
Raleigh, North Carolina 27602
(919) 733-2178

North Dakota
Department of Health
Division of Hazardous Waste Management and Special Studies
1200 Missouri Avenue
Bismarck, North Dakota 58502-5520
(701) 224-2366

Northern Mariana Islands Commonwealth of
Department of Environmental and Health Services
Division of Environmental Quality
P.O. Box 1304
Saipan, Commonwealth of Mariana Islands 96950
Overseas call: (670)234-6984

Ohio
Ohio EPA
Division of Solid and Hazardous Waste Management
361 East Broad Street
Columbus, Ohio 43266-0558
(614) 466-7220

Oklahoma
Waste Management Service
Oklahoma State Department of Health
P.O. Box 53551
Oklahoma City, Oklahoma 73152
(405) 271-5338

Oregon
Hazardous and Solid Waste Division
P.O. Box 1760
Portland, Oregon 97207
(503) 229-6534
Toll Free: (800) 452-4011

Pennsylvania
Bureau of Waste Management
Division of Compliance Monitoring
P.O. Box 2063
Harrisburg, Pennsylvania 17120
(717) 787-6239

Puerto Rico
Environmental Quality Board
P.O. Box 11488
Santurce, Puerto Rico 00910-1488
(809) 723-8184
(809) 723-8148
-or-
EPA Region II
Air and Waste Management Div.
26 Federal Plaza
New York, New York 10278
(212) 264-5175

Rhode Island
Department of Environmental Management
Division of Air and Hazardous Materials
Cannon Building, 75 Davis Street
Providence, Rhode Island 02908
(401) 277-2797

South Carolina
Department of Health and Environmental Control
Bureau of Solid and Hazardous Waste Management
2600 Bull Street
Columbia, South Carolina 29201
(803) 734-5200

South Dakota
Department of Water and Natural Resources
Office of Air Quality and Solid Waste
Foss Building, Room 217
Pierre, South Dakota 57501
(605) 773-3153

Tennessee
Division of Solid Waste Management
Tennessee Department of Public Health
701 Broadway
Nashville, Tennessee 37219-5403
(615) 741-3424

State Hazardous Waste Management Agencies

Texas
Texas Water Commission
Hazardous and Solid Waste Division
Attn: Program Support Section
1700 North Congress
Austin, Texas 78711
(512) 463-7761

Utah
Department of Health
Bureau of Solid and Hazardous Waste Management
P.O. Box 16700
Salt Lake City, Utah 84116-0700
(801) 538-6170

Vermont
Agency of Environmental Conservation
103 South Main Street
Waterbury, Vermont 05676
(802) 244-8702

Virgin Islands
Department of Conservation and Cultural Affairs
P.O. Box 4399
Charlotte Amalie, St. Thomas
Virgin Islands 00801
(809) 774-3320
-or-
EPA Region II
Air and Waste Management Division
26 Federal Plaza
New York, New York 10278
(212) 264-5175

Virginia
Department of Health
Division of Solid and Hazardous Waste Management
Monroe Building, 11th Floor
101 North 14th Street
Richmond, Virginia 23219
(804) 225-2667
Hazardous Waste Hotline: (800) 552-2075

Washington
Department of Ecology
Solid and Hazardous Waste Program
Mail Stop PV-11
Olympia, Washington 98504-8711
(206) 459-6322
In-State: 1-800-633-7585

West Virginia
Division of Water Resources
Solid and Hazardous Waste/Ground Water Branch
1201 Greenbrier Street
Charleston, West Virginia 25311

Wisconsin
Department of Natural Resources
Bureau of Solid Waste Management
P.O. Box 7921
Madison, Wisconsin 53707
(608) 266-1327

Wyoming
Department of Environmental Quality
Solid Waste Management Program
122 West 25th Street
Cheyenne, Wyoming 82002
(307) 777-7752
-or-
EPA Region VIII
Waste Management Division (8HWM-ON)
One Denver Place
999 18th Street
Suite 1300
Denver, Colorado 80202-2413
(303) 293-1502

Department of Transportation Shipping Codes

DOT Shipping Names	Hazard Class	UN or NA #
Acetic acid, glacial (RQ-1000/4540)	Corrosive material	UN2789
Acetone	Flammable liquid	UN1090
Acetylene	Flammable gas	UN1001
Acrylic acid	Corrosive material	UN2218
Adhesive	Combustible liquid	UN1133
Adhesive	Flammable liquid	UN1133
Alkaline (corrosive) liquid, n.o.s.	Corrosive material	NA1719
Aluminum chloride anhydrous	Corrosive material	UN1726
Ammonia, anhydrous (RQ-100/45.4)	Nonflammable gas	UN1005
Ammonium hydroxide (containing not less than 12% but not more than 44% ammonia (RQ-1000/454)	Corrosive material	NA2672
Ammonium nitrate mixed fertilizer	Oxidizer	UN2069
Asbestos	ORM-C	
Battery, electric storage, wet filled with acid	Corrosive material	UN2794
Asphalt, at or above its flashpoint	ORM-C	NA1999
Asphalt, cut back	Flammable liquid	NA1999
Battery fluid, acid	Corrosive material	UN2796
Battery fluid, alkali, with battery, electric storage wet, empty or dry	Corrosive material	UN2797
Benzene (benzol) (RQ-1000/454)	Flammable liquid	UN1114
Bromine	Corrosive material	UN1744
Butyl acetate (RQ-5000/2270)	Flammable liquid	UN1123
Calcium hypochlorite, hydrated	Oxidizer	UN2880
Calcium hypochlorite mixture, dry	Oxidizer	UN1748
Calcium oxide	ORM-B	UN1910
Carbon tetrachloride (RQ-5000/2270)	ORM-A	UN1846
Cement, container, linoleum tile, or wallboard, liquid	Flammable liquid	NA1133
Cement, roofing, liquid	Flammable liquid	NA1133
Chlordane, liquid (RQ-1/0.454)	Combustible liquid	NA2762
Chlorine (RQ-10/4.54)	Nonflammable gas	UN1017
Chlorobenzene (RQ-100/45.4)	Flammable liquid	UN1134
Coal tar distillate	Combustible liquid	UN1137
Combustible liquid, n.o.s.	Combustible liquid	NA1993
Compound, cleaning, liquid	Corrosive material	NA1760
Compound, cleaning, liquid hydrochloric (muriatic) acid	Corrosive liquid	NA1789
Copper based pesticide, liquid, n.o.s. (compounds and preparations)	Flammable liquid	UN2776
Corrosive liquid, n.o.s.	Corrosive material	UN1760
Corrosive solid	Corrosive material	UN1759

Department of Transportation Shipping Codes		
DOT Shipping Names	Hazard Class	UN or NA #
Cyanide solution, n.o.s.	Poison B	UN1935
Denatured alcohol	Flammable liquid	NA1986
Disinfectant, liquid	Corrosive material	UN1903
Disinfectant, liquid	Poison B	UN1601
Endrin (RQ-1/0.454)	Poison B	NA2761
Ethyl acetate	Flammable liquid	UN1173
Ethyl alcohol	Flammable liquid	UN1170
Ethylene glycol diethyl ether	Combustible liquid	UN1153
Flammable liquid, n.o.s.	Flammable liquid	UN1993
Flammable solid, n.o.s.	Flammable solid	UN1325
Hazardous substance, liquid or solid, n.o.s.	ORM-E	NA9188
Hazardous waste, liquid or solid, n.o.s.	ORM-E	NA9189
High explosive	Class A explosive	
Hydrazine, aqueous solution	Corrosive material	UN2030
Hydrochloric acid (RQ-5000/2270)	Corrosive material	UN1789
Hydrogen peroxide solution	Oxidizer	UN2014
Hypochlorite solution containing not more than 7% available chlorine by weight (RQ-100/45.4)	ORM-B	NA1791
Insecticide, liquid, n.o.s.	Combustible liquid	NA1993
Insecticide, liquid, n.o.s.	Flammable lilquid	NA1993
Insecticide, liquid, n.o.s.	Poison B	NA2902
Isopropanol	Flammable liquid	UN1219
Kerosene	Combustible liquid	UN1223
Lime, unslaked. (See calcium oxide)	ORM	
Lindane (RQ-1/.454)		NA2761
Methyl Acetone	Flammable liquid	UN1232
Methyl alcohol	Flammable liquid	UN1230
Muriatic acid. See Hydrochloric acid		
Naphtha	Combustible liquid	UN2553
Naphthalene or Naphthalin (RQ-5000/2270)	0RM-A	UN1334
Nitrating acid, spent (RQ-1000)/454)	Corrosive material	NA1826
Oil, described as oil, Oil, n.o.s., Petroleum oil, or Petroleum oil, n.o.s.	Flammable liquid	NA1270
Paint	Flammable liquid	UN1263
Parathion, liquid (RQ-1/0.454)	Poison B	NA2783
Perchloric acid	Oxidizer	UN1873
Petroleum naphtha	Flammable liquid	UN1255
Phenol (RQ-1000/454)	Poison B	UN1671
Poisonous liquid or gas n.o.s.	Poison A	NA1955

Department of Transportation Shipping Codes		
DOT Shipping Names	Hazard Class	UN or NA #
Potassium hydroxide, dry solid, flake, bead, or granular (RQ-1000/454)	Corrosive material	UN1813
Potassium hydroxide, liquid or solution (RQ-1000/454)	Corrosive material	UN1814
Radioactive material, n.o.s.	Radioactive material	UN2982
Refrigerant gas, n.o.s.	Nonflammable gas	UN1078
Refrigerant gas, n.o.s.	Flammable gas	NA1954
Sodium hydroxide, dry solid, flake, bead, or granular (RQ-1000/454)	Corrosive material	UN1823
Sodium hydroxide, liquid or solution (RQ-1000/454)	Corrosive material	UN1824
Sodium Hypochlorite. See Hypochlorite solution		
Toluene (toluol) (RQ-1000/454)	Flammable liquid	UN1294
1,1,1 – Trichloroethane	ORM-A	UN2831
Trichloroethylene (RQ-1000/454)	ORM-A	UN1710
Trifluorochloroethylene	Flammable gas	UN1082
Turpentine	Flammable liquid	UN1299
Vinyl chloride	Flammable gas	UN1086
Xylene (xylol) (RQ-1000/454)	Flammable liquid	UN1307
Zinc bromide (RQ-5000/2270)	ORM-E	NA9156

OMB#: 2050-0024 Expires 12-31-88

BEFORE COPYING FORM, ATTACH SITE IDENTIFICATION LABEL OR ENTER:

SITE NAME ______________________

EPA ID NO. |_|_|_|_|_|_|_|_|_|_|_|_|

FORM **IC**

U.S. ENVIRONMENTAL PROTECTION AGENCY

1987 Hazardous Waste Generation and Shipment Report

IDENTIFICATION AND CERTIFICATION

WHO MUST COMPLETE THIS FORM? Form IC must be completed by every site that received this package.

INSTRUCTIONS: Please read the detailed instructions beginning on page 8 of the 1987 Hazardous Waste Generation and Shipment Report Instructions booklet before completing this form.

Complete Sections I through IV and Sections VI through IX immediately. Complete Section V, certification, after you have finished the full report package.

SEC. I. Site name and physical location which may differ from the mailing address. Complete items A through G. Mark [X] for items A, B, C, D, F, and G if same as label; if different, enter corrections. If label is absent, enter information.

A. Site/company name Same as label ☐ or ⟶			B. EPA ID No. Same as label ☐ or ⟶ |_|_|_|_|_|_|_|_|_|_|_|_|
C. Address number and street name of physical location - if not known, enter industrial park, building name or other physical location description Same as label ☐ or ⟶			
D. City, town, village, etc. Same as label ☐ or ⟶	E. County	F. State Same as label ☐ or ⟶ |_|_|	G. Zip Code Same as label ☐ or ⟶ |_|_|_|_|_|-|_|_|_|_|

SEC. II. Mailing address of site. Mark [X] for A, B, C, and D if same as label; if different, enter corrections.

A. Number and street name of mailing address Same as label ☐ or ⟶		
B. City, town, village, etc. Same as label ☐ or ⟶	C. State Same as label ☐ or ⟶ |_|_|	D. Zip Code Same as label ☐ or ⟶ |_|_|_|_|_|-|_|_|_|_|

SEC. III. Name, title, and telephone number of the person who should be contacted if questions arise regarding this report.

A. Please print: Last name	First name	M.I.	B. Title	C. Telephone |_|_|_| |_|_|_|-|_|_|_|_| Extension |_|_|_|_|

SEC. IV. Enter the Standard Industrial Classification (SIC) Code that describes the principal products, group of products, produced or distributed, or the services rendered at the site's physical location. Enter more than one SIC Code only if no one industry description includes the combined activities of the site. SIC codes are listed beginning on page 1 of the 1987 Hazardous Waste Generation, Shipment and Management Report Codebook.

A.	B.	C.	D.	E.	F.
|_|_|_|_|	|_|_|_|_|	|_|_|_|_|	|_|_|_|_|	|_|_|_|_|	|_|_|_|_|

SEC V. I certify under penalty of law that I have personally examined and am familiar with the information submitted in this and all attached documents, and that based on my inquiry of those individuals immediately responsible for obtaining the information, I believe that the submitted information is true, accurate, and complete. I am aware that there are significant penalties for submitting false information, including the possibility of fine and imprisonment.

A. Please print: Last name	First name	M.I.	Title
B. Signature			Date of signature |_|_| Mo. |_|_| Day |_|_| Yr.

Page 1 of ________

OVER ⟶

EPA Form 8700 - 13A (5-80) (Rev. 11-85) Revised (12-87)

FORM IC

SEC. VI. Does this site's EPA ID authorize hazardous waste generation?

☐ NO ——→ SKIP TO SECTION VII.

☐ YES ——→ Did this site generate any hazardous waste during 1987?

☐ YES ——→ READ DETAILED INSTRUCTION ON PAGE 9 OF THE 1987 HAZARDOUS WASTE GENERATION AND SHIPMENT REPORT INSTRUCTIONS BOOKLET FOR ACUTE AND ACCUMULATION LIMITS. MARK [X] NEXT TO THE HAZARDOUS WASTE GENERATION QUANTITY CATEGORY THAT APPLIED TO THIS SITE DURING 1987.

☐ Category 1: More than 1000 kg (2,200 lb) in one or more months
☐ Category 2: More than 100 kg (220 lb) but no more than 1000 kg (2,200 lb) in any single month
☐ Category 3: No more than 100 kg (220 lb) in any single month

☐ Mark [X] if this site changed from Category 1 to Category 2 or 3 due to waste minimization activity conducted during 1986 or 1987.

☐ NO ——→ CONTINUE BELOW, MARK [X] NEXT TO ALL THAT APPLY.

☐ Generated, excluded or delisted wastes
☐ Generated hazardous waste prior to 1987 but do not expect to generate in the future - MARK [X] FOR REASON IN ONE BOX BELOW

☐ Waste was from one-time event(s) (e.g. spills, remedial actions, etc.)
☐ Waste minimization activity undertaken during 1986 or 1987
☐ Out of business

☐ Generated hazardous waste prior to 1987 and expect to generate in the future
☐ Never generated before but expect to generate in the future
☐ Never generated and do not expect to generate in the future - MARK [X] FOR REASON IN ONE BOX BELOW

☐ Protective notifier only
☐ Misunderstood the requirements
☐ Notified to secure transportation services
☐ Other EXPLAIN REASON FOR GENERATOR NOTIFICATION IN COMMENTS

SEC. VII. Does this site have RCRA Interim Status or a RCRA permit to treat, store, or dispose hazardous waste?

☐ NO ——→ SKIP TO SECTION VIII

☐ YES ——→ Did the site treat, store, or dispose (T/S/D) hazardous waste in RCRA-regulated units during 1987?

☐ YES ——→ SKIP TO SECTION VIII

☐ NO ——→ CONTINUE BELOW, MARK [X] NEXT TO ALL THAT APPLY

☐ T/S/D excluded waste during 1987
☐ T/S/D hazardous waste in exempt units during 1987
☐ T/S/D hazardous waste prior to 1987 but did not T/S/D waste during 1987. MARK [X] IN ONE BOX BELOW

☐ T/S/D will resume in the future
☐ Have notified of planned closure
☐ Site is in closure or post closure

☐ Never T/S/D hazardous waste prior to 1987 but: MARK [X] IN ONE BOX BELOW

☐ Expect to T/S/D hazardous waste in the future
☐ Do not expect to T/S/D hazardous waste in the future - EXPLAIN REASON FOR INTERIM STATUS OR PERMIT IN COMMENTS

SEC. VIII. Do you wish to withdraw this site's generator notification or EPA Part A permit application?

Withdraw generator notification ☐ Yes ☐ No
Withdraw Part A permit application ☐ Yes ☐ No

SEC. IX. Does this site have an area not requiring a RCRA Part A or Part B permit that is used exclusively for the short term accumulation of hazardous waste?

☐ NO

☐ YES ——→ DOES THE AREA HAVE:

Containers ☐ No ☐ Yes
Tanks ☐ No ☐ Yes ——→ ENTER THE NUMBER OF TANKS AND THEIR TOTAL CAPACITY IN GALLONS.
|_|_|_| Number |_|_|_|_|_|_|_|_|_| Gallon capacity

Comments:

Page 2 of ________

BEFORE COPYING FORM, ATTACH SITE IDENTIFICATION LABEL OR ENTER:

SITE NAME ______________________

EPA ID NO. [ ]

U.S. ENVIRONMENTAL PROTECTION AGENCY

1987 Hazardous Waste Generation and Shipment Report

FORM GS

WASTE GENERATION AND SHIPMENT

WHO MUST COMPLETE THIS FORM? Form GS must be completed by every site that generated hazardous waste on site or shipped hazardous waste off site during 1987.

☐ Mark ☒ if you are not required to complete Form GS.

INSTRUCTIONS: Please read the detailed instructions beginning on page 12 of the 1987 Hazardous Waste Generation and Shipment Report Instructions booklet before completing this form.

Make and complete a photocopy of this form for each hazardous waste generated on site or shipped off site during 1987.

Complete Sections I through IV. Throughout this form enter "DK" if the information requested is not known or is not available; enter "NA" if the information requested is not applicable.

Sec. I

A. Waste description — Instruction Page 12

B. EPA hazardous waste code — Page 12

C. State hazardous waste code — Page 13

D. SIC code — Page 13

E. Source code — Page 13

F. Waste form code — Page 13

G. Waste minimization results — Page 13

Sec. II

A. Organics — Instruction Page 14: High, Low, Test, Note

B. Water — Page 15: High, Low, Note

C. Total Solids — Page 15: High, Low, Note

D. Suspended Solids — Page 15: High, Low, Note

E. BTU — Page 16: High, Low, UOM, Note

F. Toxic Metals — Page 16: Note

Metal	High	Low	Test
1.			
2.			
3.			
4.			
5.			
6.			

G. pH — Page 18: High, Low, Note

H. Flashpoint — Page 18: High °F, Low °F, Note

I. Cyanides — Page 19: High, Low, Test, Note

J. Halogens — Page 20: High, Low, Note

K. Radioactive — Page 20: Yes, No, Note

Sec. III

A. 1986 quantity hazardous waste generated — Instruction Page 20

B. 1987 quantity hazardous waste generated — Page 20

C. UOM — Page 21

D. Density — Page 21: ☐ lbs/gal ☐ sg

E. Quantity hazardous waste on site on January 1, 1987 — Page 21

F. Quantity hazardous waste remaining on site on December 31, 1987 — Page 21

Sec. IV

A. EPA ID No. of facility to which waste was shipped — Instruction Page 22

B. Number of shipments — Page 22

C. Transport mode — Page 22

D. Off-site T/S/D/R code — Page 22

E. Total quantity shipped — Page 22

Comments:

Page ____ of ____

BEFORE COPYING FORM, ATTACH SITE IDENTIFICATION LABEL OR ENTER:

SITE NAME ______________________

EPA ID NO. |_|_|_|_|_|_|_|_|_|_|_|_|

UNITED STATES ENVIRONMENTAL PROTECTION AGENCY

FORM

OI

U.S. ENVIRONMENTAL PROTECTION AGENCY

1987 Hazardous Waste Generation and Shipment Report

OFF-SITE IDENTIFICATION

WHO MUST COMPLETE THIS FORM? Form OI must be completed by every site that shipped hazardous waste off site and every site that received hazardous waste from off site during 1987.

☐ Mark [X] if you are not required to complete Form OI.

INSTRUCTIONS: Please read the detailed instructions beginning on page 23 of the 1987 Hazardous Waste Generation and Shipment Report Instructions booklet before completing this form.

Complete A through E for each off-site installation to which you shipped waste or from which you received waste during 1987.

Complete A through D for every transporter you used during the reporting year.

Throughout this form enter "DK" if the information requested is not known or is not available; enter "NA" if the information requested is not applicable. Make and complete additional copies of this form if you need to identify more than four off-site installations or transporters.

Site 1

A. EPA ID No. of off-site installation or transporter Instruction page 23		B. Name of off-site installation or transporter Page 23
|_|_|_|_|_|_|_|_|_|_|_|_|		
C. Site type code Page 24	D. Site relationship code Page 24	E. Address of off-site installation Page 24
|_|	|_|	Street ______ City ______ State |_|_| Zip Code |_|_|_|_|_|–|_|_|_|_|

Site 2

A. EPA ID No. of off-site installation or transporter Instruction page 23		B. Name of off-site installation or transporter Page 23
|_|_|_|_|_|_|_|_|_|_|_|_|		
C. Site type code Page 24	D. Site relationship code Page 24	E. Address of off-site installation Page 24
|_|	|_|	Street ______ City ______ State |_|_| Zip Code |_|_|_|_|_|–|_|_|_|_|

Site 3

A. EPA ID No. of off-site installation or transporter Instruction page 23		B. Name of off-site installation or transporter Page 23
|_|_|_|_|_|_|_|_|_|_|_|_|		
C. Site type code Page 24	D. Site relationship code Page 24	E. Address of off-site installation Page 24
|_|	|_|	Street ______ City ______ State |_|_| Zip Code |_|_|_|_|_|–|_|_|_|_|

Site 4

A. EPA ID No. of off-site installation or transporter Instruction page 23		B. Name of off-site installation or transporter Page 23
|_|_|_|_|_|_|_|_|_|_|_|_|		
C. Site type code Page 24	D. Site relationship code Page 24	E. Address of off-site installation Page 24
|_|	|_|	Street ______ City ______ State |_|_| Zip Code |_|_|_|_|_|–|_|_|_|_|

Comments:

Page ____ of ____

BEFORE COPYING FORM, ATTACH SITE IDENTIFICATION LABEL OR ENTER:

SITE NAME ______________________

____

EPA ID NO. |_|_|_|_|_|_|_|_|_|_|_|_|

UNITED STATES ENVIRONMENTAL PROTECTION AGENCY

FORM
WM

U.S. ENVIRONMENTAL PROTECTION AGENCY

1987 Hazardous Waste Generation and Shipment Report

WASTE MINIMIZATION

PART I

WHO MUST COMPLETE THIS FORM?

Form WM Part I, describing efforts undertaken to implement waste minimization programs, must be completed by all generators required to file an Annual/Biennial Report. This requirement was established in response to statutory provisions included in the Hazardous and Solid Waste Amendments of 1984 (HSWA).

NOTE: Generators shipping hazardous waste off site are required to certify, on Item 16 of the Uniform Hazardous Waste Manifest, that they have a program in place to reduce, to the degree determined economically practicable, the volume and toxicity of hazardous waste generated. A similar certification must also be made by generators who have obtained a RCRA treatment, storage, or disposal permit. Consistent with these certification requirements, generators must report, on Form WM Part I, the efforts undertaken to implement waste minimization programs.

INSTRUCTIONS:

Please read the detailed instructions on page 25 of the 1987 Hazardous Waste Generation and Shipment Report Instructions booklet before completing this form.

Answer questions 1 through 10. Throughout this form enter "DK" if the information requested is not known or is not available; enter "NA" if the information requested is not applicable.

1. Did this site create or expand a source reduction and recycling program?

	1987		1986		Prior Years	
	Yes	No	Yes	No	Yes	No
Create	☐	☐	☐	☐	☐	☐
Expand	☐	☐	☐	☐	☐	☐

2. Did this site have a written policy or statement that outlined goals, objectives and methods for source reduction and recycling of hazardous waste?

	1987	1986	Prior Years
Yes	☐	☐	☐
No	☐	☐	☐

3. What was the dollar amount of capital expenditures (plant and equipment) and operating costs devoted to source reduction and recycling of hazardous waste? ENTER ZERO (0) IF NONE.

	1987	1986	Prior Years
Capital expenditures	$ ________	$ ________	$ ________
Operating costs	$ ________	$ ________	$ ________

4. Did this site have an employee training program or provide incentives (bonuses, awards, personal recognition, etc.) to identify and implement source reduction and recycling opportunities and activities?

	1987		1986		Prior Years	
	Yes	No	Yes	No	Yes	No
Training	☐	☐	☐	☐	☐	☐
Incentives	☐	☐	☐	☐	☐	☐

Page ____ of ____

OVER --->

FORM WM - PART I

5. Did this site conduct a source reduction and/or recycling opportunity assessment or audit? Note: an opportunity assessment or audit is a procedure that identifies practices that can be implemented to reduce the generation of hazardous waste or the quantity which must subsequently be treated, stored or disposed.

	1987		1986		Prior Years	
	Yes	No	Yes	No	Yes	No
Site-Wide	☐	☐	☐	☐	☐	☐
Process-Specific	☐	☐	☐	☐	☐	☐

6. Did this site identify or implement new SOURCE REDUCTION opportunities to reduce the volume and/or toxicity of hazardous waste generated at this site?

	1987		1986		Prior Years	
	Yes	No	Yes	No	Yes	No
Identify	☐	☐	☐	☐	☐	☐
Implement	☐	☐	☐	☐	☐	☐

7. What factors have delayed or prevented implementation of SOURCE REDUCTION opportunities. MARK ☒ NEXT TO ALL THAT APPLY.

- ☐ a. Insufficient capital to install new source reduction equipment or implement new source reduction practices.
- ☐ b. Lack of technical information on source reduction techniques, applicable to my specific production processes.
- ☐ c. Source reduction is not economically feasible: cost savings in waste management or production will not recover the capital investment.
- ☐ d. Concern that product quality may decline as a result of source reduction.
- ☐ e. Technical limitations of the production processes.
- ☐ f. Permitting burdens.
- ☐ g. Other (SPECIFY) ______________________________

8. Did this site identify or implement new RECYCLING opportunities to reduce the volume and/or toxicity of hazardous waste generated at this site or subsequently treated, stored, or disposed of on site or off site?

	1987		1986		Prior Years	
	Yes	No	Yes	No	Yes	No
Identify	☐	☐	☐	☐	☐	☐
Implement	☐	☐	☐	☐	☐	☐

Page _____ of _____

FORM WM - PART I

EPA ID NO. |_|_|_|_|_|_|_|_|_|_|_|_|_|_|_|

9. What factors have delayed or prevented implementation of on-site or off-site RECYCLING opportunities. MARK ☒ NEXT TO ALL THAT APPLY.

- ☐ a. Insufficient capital to install new recycling equipment or implement new recycling practices.
- ☐ b. Lack of technical information on recycling techniques applicable to this site's specific production processes.
- ☐ c. Recycling is not economically feasible: cost savings in waste management or production will not recover the capital investment.
- ☐ d. Concern that product quality may decline as a result of recycling.
- ☐ e. Requirements to manifest wastes inhibit shipments off site for recycling.
- ☐ f. Financial liability provisions inhibit shipments off site for recycling.
- ☐ g. Technical limitations of product processes inhibit shipments off site for recycling.
- ☐ h. Technical limitations of production processes inhibit on-site recycling.
- ☐ i. Permitting burdens inhibit recycling.
- ☐ j. Lack of permitted off-site recycling facilities.
- ☐ k. Unable to identify a market for recyclable materials.
- ☐ l. Other (SPECIFY) ______________________________

10. Has this site requested or received technical information or financial assistance on source reduction and/or recycling practices from any of the following sources? MARK ☒ NEXT TO ALL THAT APPLY.

	1987		1986		Prior Years	
	Technical	Financial	Technical	Financial	Technical	Financial
a. Local government	☐	☐	☐	☐	☐	☐
b. State government	☐	☐	☐	☐	☐	☐
c. Federal government	☐	☐	☐	☐	☐	☐
d. Trade associations	☐	☐	☐	☐	☐	☐
e. Educational institutions	☐	☐	☐	☐	☐	☐
f. Suppliers	☐	☐	☐	☐	☐	☐
g. Other parts of your firm	☐	☐	☐	☐	☐	☐
h. Other firms/consultants	☐	☐	☐	☐	☐	☐
i. No request made	☐	☐	☐	☐	☐	☐
j. Other (conferences, literature, etc.)________________	☐	☐	☐	☐	☐	☐

Comments:

Page _____ of _____

BEFORE COPYING FORM, ATTACH SITE IDENTIFICATION LABEL OR ENTER:

SITE NAME ____________________

____

EPA ID NO. [| | | | | | | | | | | |]

UNITED STATES ENVIRONMENTAL PROTECTION AGENCY

FORM
WM

U.S. ENVIRONMENTAL PROTECTION AGENCY

1987 Hazardous Waste Generation and Shipment Report

WASTE MINIMIZATION

PART II

WHO MUST COMPLETE THIS FORM? Form WM Part II must be completed only by generators that engaged in an activity during 1987 that resulted in waste minimization.

Waste minimization means:
(1) reduction in the volume and/or toxicity of hazardous waste generated as a result of source reduction; and/or,
(2) reduction in the volume and/or toxicity of hazardous waste subsequently treated, stored, or disposed as a result of on-site or off-site recycling.

☐ Mark ☒ and do not complete this form if no waste minimization results were achieved during 1987.

INSTRUCTIONS: Please read the detailed instructions beginning on page 26 of the 1987 Hazardous Waste Generation and Shipment Report Instructions booklet before completing this form.

Make and complete a photocopy of this form for each hazardous waste minimized in 1987.

Complete Sections I through IV. Throughout this form enter "DK" if the information requested is not known or is not available; enter "NA" if the information requested is not applicable.

Sec. I

A. EPA hazardous waste code Instruction Page 27	B. State hazardous waste code Page 27	C. Product or service description Page 27	D. Product or service SIC code Page 27

E. Waste form code Page 27	F. UOM Page 28	G. Density Page 28	H. Source description: Page 28	I. Source code Page 28
		☐ • ☐ ☐ lbs/gal ☐ sg		

Sec. II

A. 1986 quantity generated Instruction Page 29	B. 1987 quantity generated Page 29	C. Production ratio Page 29	D. Toxicity change code Page 31
		☐ • ☐	

E. Waste minimization: recycling Page 31
Code 1. ☐ 2. ☐ Quantity recycled

F. Waste minimization: source reduction Page 32
Code 1. ☐ 2. ☐ 3. ☐ Quantity prevented

Sec. III

A. Narrative description of waste minimization project or activity and results achieved Instruction Page 39

Page ____ of ____

OVER —>

FORM WM - PART II

Sec. IV.	**Instructions:** Answer questions 1 through 4. Mark ☒ next to the effects produced by the source reduction and/or recycling activity reported on this form in Sections I through III.

1. What effect did this site's source reduction and/or recycling activity have on the **quantity** of **water effluent** produced by hazardous waste generation processes during 1987?

☐ a. Increase in the quantity of water effluent
☐ b. Decrease in the quantity of water effluent
☐ c. No effect on the quantity of water effluent
☐ d. Don't know

2. What effect did this site's source reduction and/or recycling activity have on the **toxicity** of **water effluent** produced by hazardous waste generation processes during 1987?

☐ a. Increase in the concentration of hazardous constituents
☐ b. Decrease in the concentration of hazardous constituents
☐ c. No effect on the concentration of hazardous constituents
☐ d. Don't know

3. What effect did this site's source reduction and/or recycling activity have on the **quantity** of **air emissions** produced by hazardous waste generation processes during 1987?

☐ a. Increase in the quantity of air emissions
☐ b. Decrease in the quantity of air emissions
☐ c. No effect on the quantity of air emissions
☐ d. Don't know

4. What effect did this site's source reduction and/or recycling activity have on the **toxicity** of the **air emissions** produced by hazardous waste generation processes during 1987?

☐ a. Increase in the concentration of hazardous constituents
☐ b. Decrease in the concentration of hazardous constituents
☐ c. No effect on the concentration of hazardous constituents
☐ d. Don't know

Comments:

Page ____ of ____

Absorbent
A substance that attracts and holds large quantities of liquid.

Acetylene
A specific gas which, when combined with pure oxygen and ignited, produces an extremely hot flame.

Acid
A liquid exhibiting a pH less than 7.0.

Acute Toxic Chemical
A substance that causes rapid, but not necessarily permanent, adverse health effects.

ACGIH
See American Conference of Governmental Industrial Hygienists.

Air Stripping
A process by which volatile compounds are removed from soil or water by blowing clean air through the soil or water.

Alkalai
A liquid exhibiting a pH greater than 7.0.

American Conference of Governmental Industrial Hygienists
An organization of professional scientists skilled in the arts of industrial hygiene.

Annual Report of Carcinogens
A list of those substances determined by the National Toxicology Program to be carcinogenic to humans.

APR
An Air Purifying Respirator, a device that removes pollutants from a contaminated atmosphere as a person breathes.

Asbestos
One of several minerals characterized by very small, fibrous shapes, which, when inhaled, usually cause lung cancer.

Asphixiant
A substance that interferes with the normal operation of the lungs.

Bi-Annual Report
A report of waste reduction and recycling activities required of all hazardous waste generators.

Berm
A raised wall enclosing a liquid waste storage or spill area.

Bioaccumulative
A substance that enters the body, but is not removed through normal bodily functions.

Carcinogen
A cancer-causing substance.

Caustic
An alkaline material.

CERCLA
See Comprehensive Environmental Response, Compensation, and Liability Act.

CFR
See Code of Federal Regulations.

Chemical Hazard Response Information System
A series of monographs prepared for the U.S. Coast Guard, each of which provides detailed hazard information on a particular chemical.

Chemical Stability
The tendency of a substance to remain unchanged over a wide range of physical, environmental, and chemical conditions.

Chloracne
A skin rash, caused by exposure to chlorine, characterized by small pimples and inflammation of skin pores.

Chlorinated Solvent
A material that dissolves greases and oils, and contains one or more chlorine atoms in its chemical structure.

Chlorine
A highly corrosive elemental gas.

CHRIS
See Chemical Hazard Response Information System.

Chronic Toxic Chemical
A substance that causes adverse health effects which are delayed in their appearance, but usually cause permanent damage.

Citizen Suit
A legal action against EPA designed to force this organization to carry out actions to which it is obligated by statute or regulation.

Class Ia Flammable Liquid
A liquid having a flash point below 73°F (22.8°C) and a boiling point below 100°F (37.8 °C).

Class Ib Flammable Liquid
A liquid having a flash point below 73°F (22.8°C) and a boiling point at or above 100°F (37.8°C).

Class Ic Flammable Liquid
A liquid having a flash point at or above 73°F (22.8°C) and below 100°F (37.8°C).

Class II Combustible Liquid
A liquid having a flash point at or above 100°F (37.8°C).

Class IIIa Combustible Liquid
A liquid having a flash point at or above 140°F (60°C) and below 200°F (93.4°C).

Class IIIb Combustible Liquid
A liquid having a flash point at or above 200°F (93°C).

Code of Federal Regulations
A uniform system of listing all the regulations promulgated by any federal agency.

Combustible
Able to be ignited and burned.

Compatible Wastes
Wastes which, when mixed together, will not cause an adverse chemical reaction.

Comprehensive Environmental Response, Compensation, and Liability Act
The federal law defining the government actions in the event of an uncontrolled release of hazardous materials to the environment.

Containment Box
A device to retain liquid spills. (See Figure 6.3.)

Coolant
A chemical mixture designed to rapidly remove heat from designated areas, such as automotive radiators and refrigerators.

Confidentiality Agreement
A legal document that allows the release of trade secret information to physicians and others who may need this information to protect workers from danger or to treat an exposed worker.

Construction Advisory Committee
A special committee established by Congress to advise the Secretary of Labor on matters affecting the health and safety of construction workers.

Consumer Products Safety Act
The federal law that establishes standards of safety for consumer products.

Corrosive
A substance that causes destruction of human tissue or metal on contact.

Cyanosis
A chemical reaction in the blood which prevents the transfer of oxygen to red blood cells.

Department of Health and Human Services
The federal agency responsible for establishing health and safety standards for the protection of persons.

Department of Labor
The federal agency that oversees all laws associated with hiring, employing, and protecting workers.

Department of Transportation
The federal agency responsible for regulating the safe transportation of goods and materials within the United States.

Dermatitis
An inflammation of the skin.

Dessicant
A substance that rapidly absorbs water from air.

D.O.T.
See Department of Transportation.

D.O.T. Shipping Name
A material classification name assigned by D.O.T. to all transportable materials. (See Appendix D.)

Drip Pan
A device for catching and containing drips and leaks from liquid storage drums. (See Figures 6.5 and 6.6.)

Electrical Conductivity
The ability to transmit an electrical current.

Electrolyte
A substance which breaks down into ions in solution or when fused, thereby becoming electrically conductive.

Encapsulation
A technique for trapping asbestos fibers in a dense chemical substance.

Environmental Protection Agency
The federal agency responsible for providing environmental safety.

EPA
See Environmental Protection Agency.

EPA Identification Number
A unique number assigned to each generator of hazardous waste who notifies EPA of his activity.

EPA Regional Administrator
The senior EPA official in charge of EPA activities at a generation site.

Etiologic Agent
A substance which causes a disease (such as a flu virus).

Exception Report
A report sent to EPA by a generator of hazardous waste. The Exception Report is used when the generator does not receive copies of hazardous waste manifest forms from transporters or disposers of wastes which the generator has shipped off-site for disposal.

Extraction Procedure Toxicity
The propensity of a substance to emit toxins under a specific test procedure, defined by EPA, which uses an acid to extract the toxins from the substance in question.

Facility Audit
A review of activities at a facility or location, the purpose of which is to identify the improper storage, handling, or disposal of hazardous materials or waste.

Federal Hazardous Substances Act
The federal law that controls the use of hazardous constituents in consumer products.

FID
See Flame Ionization Detector.

Flammable
Easily ignited and capable of burning quickly.

Flame Ionization Detector
An ionization detector that uses hydrogen flame to generate ions.

Flash Point
The temperature at which a substance will spontaneously ignite.

Friable
Dry and easily crumbled between the fingers.

Full-face Respirator
A respirator that covers the whole face with a protective shield.

Generator
A person, firm, or entity whose activities create a hazardous waste.

Groundwater
Naturally occurring water that moves through the earth's crust, usually at a depth of several feet to several hundred feet below the earth's surface.

Half-face Respirator
A respirator that covers only the nose and mouth.

Hazard Class
A D.O.T. shipping designation code. (See Appendix D.)

Hazardous Waste
A material defined by any of several statutes and regulations, usually characterized by a propensity to cause an adverse health effect to humans.

Heavy Metal
A naturally occurring elemental metal with a high molecular weight.

Hematopoietic System
A system in the human body which produces or carries blood or blood constituents.

HEPA
High Efficiency Particulate Air.

HEPA-filter
High Efficiency Particulate Air Filter.

HEPA-filtered Vacuum
A vacuum device fitted with a high efficiency particulate removal system.

Hepatotoxin
A substance that causes liver damage.

Herbicide
A substance that kills plants on contact.

Hydrogeologic Testing
A means of determining the structure and characteristics of subsurface soils and rocks, and the way water flows through them.

IARC
See International Agency for Research on Cancer.

IARC Monograph
A brief summary of the carcinogenic effects of a particular substance as determined by the International Agency for Cancer Research.

Indicator Pollutant
An easily measured pollutant which may or may not be hazardous in a normally found concentration, but which may indicate the presence of a more dangerous pollutant.

Insecticide
A substance that kills insects.

In-situ
Occurring in place.

International Agency for Research on Cancer
An international organization established to conduct research on suspected carcinogens and their effects on humans.

Inventory Control
A management plan designed to minimize the number and quantity of hazardous substances on a construction project.

Ion
An atom with one or more electrons missing.

Ionization Detector
A device that indicates the presence of specific gaseous compounds in air by subjecting the compound to ultraviolet light or hydrogen flame, and recording the number of ions created.

Irritant
A substance that causes discomfort, such as tearing, choking, vomiting, rashes, reddening of the skin, itching, or other topical responses.

Isobar
A line drawn on a map to indicate the limits of equal contaminant concentrations.

Landfill
An engineered disposal system characterized by the burial of wastes therein.

Land Treatment Area
A defined parcel of land on which wastes are deposited for the purpose of allowing natural cleansing actions to occur.

Large Quantity Generator
A classification of the Resource Conservation and Recovery Act (RCRA). (See Figure 1.1.)

Leaching
The movement of free water out of a disposed substance and through the soil.

LUST
A leaking underground storage tank.

Manifest
A specific form used by a generator of hazardous waste to track the waste from the site of generation to the site of final treatment or disposal. (See Figure 8.5.)

Material Safety Data Sheet
A form published by manufacturers of hazardous materials to describe the hazards thereof.

MSDS
See Material Safety Data Sheet.

National Institute for Occupational Safety and Health
A separate branch of OSHA established to research ways of improving worker safety.

NA number
A classification code assigned to a particular material by the Department of Transportation. (See Appendix D.)

National Institute of Occupational Safety and Health
A federal research organization dedicated to the improvement of respirators and other worker safety equipment.

National Toxicology Program
A federal research program designed to identify the hazards associated with various chemicals and other substances.

Neoprene
A specific type of rubber compound.

Nephrotoxin
A substance that causes kidney damage.

Neutralize
To reduce the pH of an alkalai, or to raise the pH of an acid, to approximately 7.0.

Neurotoxin
A substance that adversely affects the nervous system or brain.

NIOSH
National Institute of Occupational Safety and Health.

Occupational Safety and Health Act
The federal law governing the safety of workers in the workplace.

Occupational Safety and Health Administration
The federal agency responsible for worker health and safety.

Oil/Water Separator
A device that allows oils mixed with water to become trapped in a holding section for removal, while the water is allowed to pass through for disposal.

Operator
A person or entity having direct managerial control over a hazardous waste storage, treatment, or disposal facility.

OSHA
The Occupational Safety and Health Administration or the Occupational Safety and Health Act.

Oxidizer
An agent which, when acting on another substance, causes the attachment of an oxygen atom thereto.

PCB
A polychlorinated biphenon compound.

pH
A measure of the relative acidity or alkalinity of a liquid.

Phenolic Compounds
Substances containing phenols.

Photoionization Detector
An ionization detector that uses ultraviolet light to generate ions.

PID
A photoionization detector.

Plume
An identifiable and definable stream of pollutants in an otherwise clean volume of air or water.

PPE
See Personal Protective Equipment. (See also Figure 5.1.)

Pouring Box
A device designed to contain spills that may occur when transferring liquids from one container to another. (See Figure 6.4.)

Pyrophoric
Liable to ignite or burn.

Raw Material
A material used for the construction or manufacture of goods or products.

RCRA
See Resource Conservation and Recovery Act.

Reactivity
The propensity of one substance to chemically react with another.

Reducer
A substance that chemically reacts with another by removing an oxygen atom from the chemical structure of the second substance.

Regulatory
Related to provisions specified in a regulation implementing a law.

Reproductive Toxin
A substance that chemically or biologically interferes with the reproductive process; a teratogen.

Resource Conservation and Recovery Act
The federal law controlling solid and hazardous waste disposal.

Respirator
A device used to facilitate breathing in a hazardous atmosphere.

Right-to-Know
A right granted to workers by the Occupational Safety and Health Act, by which they must be informed of the risks and hazards associated with the chemicals and substances which they are required to use in the workplace.

SAR
A Supplied Air Respirator, connected by a long air hose either to a source of compressed air or to an air pump.

SCBA
A Self-Contained Breathing Apparatus utilizing an air tank connected directly to a respirator face mask.

Sensitizer
A substance that causes a person to become susceptible to the adverse health effects of the same or different substance to which the person had previously not been susceptible.

Site Audit
A review of activities at a construction site, the purpose of which is to identify inappropriate material and waste management practices.

Site Safety Officer
The person responsible for establishing the appropriate health and safety equipment to be used by workers at a hazardous waste site.

Sludge
A solid or semi-solid residue from a chemical or biological process.

Small Quantity Generator
A classification of the RCRA. (See Figure 1.1.)

Solid Waste Disposal Act
A predecessor law to the federal Resource Conservation and Recovery Act.

SSO
A Site Safety Officer.

The Standard
The Hazard Communication Standard promulgated by OSHA.

Statutory
Related to provisions specified in a law.

Steam Stripping
A process of removing volatile contaminants from soil or water by passing steam through the soil or water.

Subsurface Contamination
The presence of hazardous materials in the soil or groundwater under a site.

Superfund
The Comprehensive Environmental Response, Compensation and Liability Act.

Surface Impoundment
A lagoon or pond designed to hold waste materials and prevent their escape to the environment.

Teratogen
A substance that adversely affects any of the human reproductive system organs or that causes malformation of the fetus.

Test Pit
A hole excavated into the soil with a backhoe or shovel for the purpose of identifying the subsurface conditions at that location.

Threshold Limit Value
The concentration of a hazardous substance to which a person may be exposed continuously without adverse health effects.

TLV
See Threshold Limit Value.

Toxic
Causing an adverse health effect.

Toxic Agent
A substance that causes an adverse health effect in humans.

TPH Test
A test used to determine the total quantity of petroleum hydrocarbons in a sample of soil or water.

Trade Secrets
Chemical formulae or manufacturing processes that enable a particular business entity to gain a competitive edge in its marketplace, the disclosure of which would cause the business to lose that advantage.

Transformer
An electrical device used to convert high voltage direct current to lower voltage alternating current.

UN Number
A classification code assigned to a particular material by D.O.T. (See Appendix D.)

Vapor Pressure
A measure of the propensity of a liquid to evaporate. This measurement is based on the rate of evaporation of water.

Very Small Quantity Generator
A classification of RCRA. (See Figure 1.1.)

Vitrification
A process of melting soil or wastes to a liquid form, and then cooling the liquid to a hard, glass-like material.

VOC
See Volatile Organic Compound.

Volatile Organic Compound
A substance containing carbon, hydrogen, and oxygen atoms, characterized by a relatively low vapor pressure at ambient air temperature.

Waste Management
Activities undertaken to minimize, contain, control, store, transport, treat, or dispose a waste material.

Waste Oil
Lubricants, typically automotive lubricants, that are no longer useable for their intended purpose.

Waste Pile
An accumulation of solid waste materials at a specific location.

Waste Stream
A continuously or regularly generated quantity of waste materials from a particular facility, process, or activity.

Water Reactive
A substance that generates heat or gas in the presence of water.

Index

D

H

R

S

T

U

V

W

X